Lecture Notes in Mathematics 2249

More information about this series at http://www.springer.com/series/304

Alexander Komech • Anatoli Merzon

Stationary Diffraction by Wedges

Method of Automorphic Functions on Complex Characteristics

 Springer

Alexander Komech
Faculty of Mathematics
University of Vienna
Vienna, Austria

Anatoli Merzon
Instituto de Fisica y Matematicas
Universidad Michoacana de San Nicolas de Hidalgo
Morelia
Michoacán, Mexico

ISSN 0075-8434   ISSN 1617-9692   (electronic)
Lecture Notes in Mathematics
ISBN 978-3-030-26698-1   ISBN 978-3-030-26699-8   (eBook)
https://doi.org/10.1007/978-3-030-26699-8

Mathematics Subject Classification (2010): Primary: 35J25, 78A45; Secondary: 35Q60

This Springer imprint is published by the registered company Springer Nature Switzerland AG.
The registered company address is: Gewerbestrasse 11, 6330 Cham, Switzerland

To the blessed memory of Nina Ilina

Preface

We present a complete solution to the classical problem of stationary diffraction by wedges with general boundary conditions (b.c.). For the Dirichlet and Neumann b.c., the solution was found by Sommerfeld in 1896 and for the impedance b.c. (or Leontovich and Robin b.c.) by Malyuzhinetz in 1958. Our approach relies on a novel "method of automorphic functions (MAF) on complex characteristics" which gives all solutions to boundary problems for second-order elliptic operators with general boundary conditions. This method is also applicable to the problems of guided water waves on a sloping beach, scattering of seismic waves, etc.

This method was introduced by one of the authors in 1973 for angles $\Phi < \pi$, and was extended in 1992–2007 to angles $\Phi > \pi$ by both authors in collaboration. The method relies on the complex Fourier-Laplace transform in two variables which in turn leads to a system of algebraic equations on the Riemann surface of complex characteristics. This system is reduced to one algebraic equation with two unknown functions on the Riemann surface. We reduce this undetermined algebraic equation to the Riemann–Hilbert problem on the Riemann surface applying Malyshev's method of automorphic functions. This reduction is the key step of our approach. Finally, the Riemann–Hilbert problem is solved in quadratures. This method generalizes the Malyuzhinetz approach introduced in the framework of the impedance boundary condition.

Our presentation contains many important details and results, which we publish here for the first time. All proofs and constructions are considerably streamlined and simplified.

We also outline the creation of the diffraction theory by Fresnel, Kirchhoff, Poincaré, and Sommerfeld and survey subsequent results by Sobolev, Malyuzhinetz, Keller, Maz'ya, Grisvard, and others on the diffraction by wedges and on related problems in angular domains.

Vienna, Austria   Alexander Komech
Morelia, Mexico   Anatoli Merzon

Contents

Chapter 1
Introduction

The diffraction by wedges plays a crucial role in radar and sonar detection, because the radars and sonars catch exclusively diffracted waves from sharp edges of the wings and never catch the reflected waves from smooth surfaces. The diffraction by wedges is described by boundary value problems in angles with various boundary conditions. Similar problems also arise in guided water waves on sloping beach, in scattering of seismic waves, etc.

1.1 Early Theory of Diffraction

The first account of the diffraction is due to Leonardo da Vinci (1452–1519). Grimaldi observed systematically the diffraction of light about 1660 and coined the word "diffraction". He observed coloured fringes on the light/shadow border in a darkened room.

The mathematical ground for the wave theory of light was built by Huygens in 1678–1690 in his theory of secondary waves. Young explained for the first time diffraction patterns (about 1802) by the interference of the secondary waves issued by *edges of an aperture*.

The first universal mathematical theory of diffraction was created by Fresnel in 1814–1827 who developed significantly the ideas of Huygens and Young taking into account the secondary waves issued by *all points of the aperture*.

This theory was justified by Kirchhoff in 1883 in the framework of the Maxwell theory of light assuming that the Cauchy data of the light field on the aperture can be calculated from incident wave. We outline details of Fresnel and Kirchhof's theory of diffraction in Chap. 3. In particular, we present the calculation of the diffraction by the half-plane in the Fresnel-Kirchhoff approximation [11, 32, 133].

Many interesting details on the history of the wave theory of light can be found in [15, 29, 75, 144].

© Springer Nature Switzerland AG 2019
A. Komech, A. Merzon, *Stationary Diffraction by Wedges*, Lecture Notes in Mathematics 2249, https://doi.org/10.1007/978-3-030-26699-8_1

1.2 Diffraction by Wedges

In 1892–1897 Poincaré developed the theory of polarization in the diffraction of light using separation of variables in 2D Helmholtz equation [113, 114].

The first *exact solution* in stationary problem of diffraction by a half-plane was calculated by Sommerfeld in 1896 in his famous paper [100, 129] inspired by some ideas of Hankel, Klein and Poincaré. Later Sommerfeld compared this exact result with the Fresnel-Kirchhoff approximation, see [133]: the agreement of the diffraction patterns on the light/shadow border was amazingly perfect at large distances from the screen! We present in Chap. 5 an update version of the Sommerfeld theory.

The stationary theory of diffraction by wedges and of related boundary problems for Dirichlet, Neumann and Robin boundary conditions was developed after the first Sommerfeld work of 1896 by many authors: by Sommerfeld himself, and also by Carslaw, Smirnov, Sobolev, Keller, Kay, Oberhettinger, Borovikov, Shilov, Maz'ya, Filippov, Grisvard, Eskin, Meister, and many others. We survey these results in Chap. 6.

The decisive breakthrough was made in 1958 by Malyuzhinetz [76, 78], who solved for the first time the stationary diffraction problem by wedges with the impedance b.c. This b.c. is of crucial importance in the stationary diffraction theory, since it corresponds to the Leontovich b.c. (6.4) in the time-dependent diffraction [72].

The Malyuzhinetz approach relies on the Sommerfeld integral representation for solutions. The key idea of Malyuzhinetz's method is an ingenious reduction to a functional equation which is equivalent to the Riemann–Hilbert problem. An update presentation of the Malyuzhinetz's method can be found in [4].

However, some central questions remain open until now. In particular,

1. The problem of diffraction by wedges was not solved under general boundary conditions.
2. The solutions, as calculated in the Sommerfeld papers and in all subsequent works on the diffraction by wedge, are some particular "standing solutions" to the wave equations. However, their distinguished role as *the limiting amplitudes* was never justified.

These problems remained open for more than 100 years and were solved in the last decades by introduction of the novel "method of automorphic functions on complex characteristics" (MAF). We present this method here for the first time with all details.

1.3   Method of Automorphic Functions on Complex Characteristics (MAF)

The MAF method was introduced in 1973 by one of the authors [52, 53] for second-order elliptic equations in a convex angle $Q \subset \mathbb{R}^2$ of magnitude $\Phi < \pi$

$$A(\partial)u(x) = \sum_{|\alpha| \leq 2} a_\alpha \partial_x^\alpha u(x) = 0, \quad x \in Q, \tag{1.1}$$

with general boundary conditions

$$B_l(\partial)u(x) = \sum_{|\alpha| \leq m_l} b_{l\alpha} \partial_x^\alpha u(x) = f_l(x), \qquad x \in \Gamma_l, \quad l = 1, 2, \tag{1.2}$$

where Γ_l are the sides of the angle (a concise survey can be found in [64]). The method gives all solutions from the space of tempered distributions for boundary operators B_l satisfying the Shapiro-Lopatinski condition. In 1992–2007 the method was extended by the authors to non-convex angle of magnitude $\Phi > \pi$, see [58, 61].

The method relies on the complex Fourier-Laplace transform in two variables, which reduces the problem to a system of algebraic equations on the Riemann surface of complex characteristics for the Cauchy data of a solution. The system is reduced to the Riemann-Hilbert problem on the Riemann surface by a development of Malyshev's method of automorphic functions. This Riemann-Hilbert problem is solved in quadratures.

This method solves the stationary diffraction problem by wedges with general boundary conditions. The method was applied by the authors in 1996–2019 to time-dependent diffraction by wedges with the Dirichlet and Neumann boundary conditions, see the next section. Our approach is applicable to general differential boundary conditions and also to *nonlocal boundary conditions* arising in the diffraction by wedges with ferromagnetic coating.

We present with details this method and its extensions in Part II of this book (Chaps. 7–20). Chapters 7–19 concern the case $\Phi < \pi$. In Chap. 20 we extend the method to angles $\Phi > \pi$. Our presentation relies on our publications [52–67] and [31, 87–98] and also contains many important details and results, which we publish here for the first time. In particular, we added new results on distributional solutions in Chap. 8. In particular, Lemma 8.3 is new. All proofs and constructions are considerably streamlined and simplified.

Remark 1.1 Our method MAF is applicable, in particular, to the Helmholtz operator $A(\partial) = \Delta + \omega^2$ which plays the central role in the diffraction theory. Let us stress however, that this method is also applicable to more general elliptic operators (1.1) with complex coefficients.

1.4   Applications of the MAF Method

This method MAF was applied to a series of problems in angles.

  I. **Time-dependent diffraction by wedges.** In [31, 90–98] and in [37, 38] we
     considered time-dependent diffraction of plane harmonic waves by wedges
     with the Dirichlet and Neumann b.c. Our main results for the time-dependent
     diffraction are as follows:

     1. In [60, 92] we proved for the first time the existence and uniqueness of
        solutions in appropriate functional classes, Limiting Amplitude Principle
        (LAP) for harmonic incident plane waves with the Dirichlet and Neumann
        b.c. and calculated the corresponding limiting amplitudes. In [66] we
        extended these results to non-smooth and aperiodic incident plane waves.

        Using these formulas we have proved for the first time in [67] that
        the Sommerfeld solution [100, 129] is the limiting amplitude for time-
        dependent diffraction by the half-plane of incident plane waves (4.4) with a
        broad class of profile functions (4.5), see Sect. 5.9.
     2. In [62, 97] we have justified the formulas obtained by Sobolev, Keller,
        Blank and Kay for the scattering of the plane pulse by wedges [50, 51,
        124, 125]. Namely, we proved the coincidence of these solutions with the
        corresponding limiting amplitudes. This coincidence implies the stability of
        these particular solutions under perturbation of the profile of incident plane
        wave.
     3. In [24, 98] we calculated the long-time asymptotics of the diffracted wave
        near the wave front.

 II. **Justification of the Sommerfeld integral representation.** The Sommerfeld
     integral representation was very productive and proved widely useful in
     diffraction by wedges. However, its universality was not justified for about 100
     years. In [65] we have justified this integral representation for any tempered
     solution to the Helmholtz equation (4.18) with $\mathrm{Im}\,\omega \neq 0$ in plane angles
     expressing the integrand via the Cauchy data of solution on one side of the
     angle. This representation plays the central role in our calculation of limiting
     amplitudes [38, 60, 92] and in our justification of the completeness of Ursell's
     trapped modes on a sloping beach [63, 89, 93].
III. **Ursell's trapped modes.** The MAF method was applied to trapped modes on
     a sloping beach, as was introduced by Ursell [139]:

     1. In [63, 64, 89] the completeness of Ursell's trapped modes is proved for the
        first time. (The first mode was discovered by Stokes [135]. The problem
        of completeness was intensively studied by Peters [111], Roseau [116],
        Lehman and Lewy [73], see also [39] and [70].)
     2. In [93] the existence and stability of Ursell's trapped modes is established
        for two-layer fluid under small perturbations of the upper layer density. The

explicit formulas for the leading and correction terms are obtained from the Sommerfeld-type representation, which we justify by the MAF.

IV. **The existence of solution to the Neumann boundary value problem and its representation in the form of the Sommerfeld integral.** In [149] the existence of the solution in the Sobolev space $H^1(Q)$ and its representation in the form of the Sommerfeld integral was proved for the Helmholtz equation (4.18) with Im $\omega \neq 0$ under the Neumann boundary conditions. This problem was solved in [149] for arbitrary angles $\Phi \in (0, 2\pi)$ for the first time. Previously this problem was solved only for the right angle [84]. Note that the uniqueness of solution to this problem was proved in [21].

We have improved this result in [96] proving that the solution belongs to $H^s(Q)$ with $s = 1 + \varepsilon$ where $\varepsilon \in [0, 1/2)$. The known Sommerfeld representation does not help in this case, since the boundary data belong to the space $H^{-1/2+\varepsilon}(\mathbb{R}^+)$ (which is proved in [83]), and in this case the Sommerfeld integral converges slowly. We used for the proof the representation of the solution via four Cauchy data (8.15), which converges faster.

V. **Fredholmness of boundary value problems with wedges.** The method gives an efficient criterion for the Fredholmness of the boundary value problems on manifolds with wedges [52, 53].

1.5   General Scheme of the MAF Method

Let us sketch the main strategy of our approach. First we consider convex angles of magnitude $\Phi < \pi$, and afterward we outline the required modifications for $\Phi > \pi$. We assume that the operator $A(\partial)$ is strongly elliptic, i.e., its *complete symbol*

$$\tilde{A}(z) := \sum_{|\alpha| \leq 2} a_\alpha (-iz)^\alpha \tag{1.3}$$

does not vanish on the real plane:

$$|\tilde{A}(z)| \geq \varkappa(|z|^2 + 1), \qquad z \in \mathbb{R}^2. \tag{1.4}$$

Moreover, we assume that the boundary operators B_l satisfy the Shapiro-Lopatinski conditions (7.8). In particular, all these conditions hold for the Helmholtz operator $A = \Delta + \omega^2$ with Im $\omega \neq 0$ in the case of the Dirichlet or Neumann b.c.

We look for solutions $u(x)$ of two types: the first ones are the "distributional solutions", which are restrictions to Q of tempered distributions on the plane $\mathbb{R}^2$, and the second ones are the "regular solutions" from the Sobolev space $H^s(Q)$ with $s > 3/2$.

I. Extension to the Plane   The first step of the method is the expression of the solution $u(x)$ via the Cauchy data on the sides Γ_l with $l = 1, 2$. Namely, the angle Q

coincides with the first quadrant $K := \mathbb{R}^+ \times \mathbb{R}^+$ in a suitable system of coordinates since $\Phi < \pi$. Then the Cauchy data are defined by

$$\begin{cases} u_1^\beta(x_1) := \partial_2^\beta u_0(x_1, 0+), & x_1 > 0 \\ u_2^\beta(x_2) := \partial_1^\beta u_0(0+, x_2), & x_2 > 0 \end{cases}, \quad \beta = 0, 1. \tag{1.5}$$

We prove that the limits (1.5) exist in the sense of tempered distributions for any distributional solution to (1.1). We denote by u_0 an extension of the solution by zero outside K, and obtain from (1.1) the equation

$$Au_0(x) = \gamma(x), \qquad x \in \mathbb{R}^2, \tag{1.6}$$

where $A = A(\partial)$ and $\gamma(x)$ is the tempered distribution, which is a sum of multiple layers,

$$\gamma(x) = \sum_{0 \le k \le N} \gamma_1^k(x_1)\delta^{(k)}(x_2) + \sum_{0 \le k \le N} \gamma_2^k(x_2)\delta^{(k)}(x_1), \qquad x \in \mathbb{R}^2.$$

Here $\gamma_l^k(x_l)$ are finite linear combinations of the derivatives $\partial^\alpha u|_{\Gamma_l}$ with $|\alpha| \le N$. The Cauchy-Kowalevski method allows us to reduce N to $N = 1$, and hence, $\gamma_l^k(x_l)$ become finite linear combinations of the Cauchy data (1.5).

Applying the Fourier transform in both variables to (1.6), we obtain the identity

$$\tilde{A}(z)\tilde{u}_0(z) = \tilde{\gamma}(z), \qquad z \in \mathbb{R}^2. \tag{1.7}$$

Hence, using the strong ellipticity (1.4), we can divide by $\tilde{A}(z)$ and obtain the formula for the solution to the problem (1.1), (1.2),

$$u_0(x) = F^{-1}\frac{\tilde{\gamma}(z)}{\tilde{A}(z)}, \qquad x \in \mathbb{R}^2; \qquad u(x) = u_0(x), \quad x \in K. \tag{1.8}$$

Thus, to find the solution, it remains to calculate the distribution $\gamma(x)$ in terms of the boundary data f_l. This calculation is the general goal of our approach.

II. Reduction to Algebraic Equations on the Riemann Surface The boundary conditions (1.2) in the Fourier-Laplace transform produce two algebraic equations for the functions $v_{l\beta}(z_l) := Fv_l^\beta$, where $v_l^\beta(x_l)$ denote the functions (or tempered distributions) $u_l^\beta(x_l)$ extended by zero for negative x_l.

Next crucial step is the derivation of the third algebraic "Connection Equation" on the Riemann surface. This equation follows from (1.7) by analytic continuation into the complex tube domain

$$\mathbb{C}K^* := \{z \in \mathbb{C}^2 : \operatorname{Im} z_1 > 0, \operatorname{Im} z_2 > 0\}. \tag{1.9}$$

Namely, both functions (or tempered distributions), $\tilde{u}_0(z)$ and $\tilde{\gamma}(z)$, can be extended to analytic functions in $\mathbb{C}K^*$ by the Paley-Wiener theorem [54, 115, 122], and the relation (1.7) also holds in $\mathbb{C}K^*$. However, the symbol $\tilde{A}(z)$ vanishes on the Riemann surface V of complex characteristics of the elliptic operator $A(\partial)$:

$$V := \{z \in \mathbb{C}^2 : \tilde{A}(z) = 0\}. \tag{1.10}$$

Hence, (1.7) with $z \in V$ implies the "Connection Equation"

$$\tilde{\gamma}(z) = 0, \qquad z \in V^* := V \cap \mathbb{C}K^*. \tag{1.11}$$

This is the third algebaric relation between the functions $v_{l\beta}(z_l)$.

This Connection Equation generalizes well-known relations between the Cauchy data on real characteristics of hyperbolic equations. A part of these equations was found by Shilov [123] and Sobolev [128]. This equation follows from the Paley-Wiener theorem in the case $\Phi < \pi$. On the other hand, for $\Phi > \pi$ this argument requires a significant modification, because in this setting the tubular domain $\mathbb{C}K^*$ is an *empty set*! Suitable modification were carried out in [58, 61], see item **VII** below.

The key role of complex zeros of symbol (1.10) was suggested by the theory of boundary value problems in the half-plane, where solutions are expressed via the "stable roots" of the characteristic equation [3], see Remark 10.4.

III. On the Equivalence of the Reduction   Thus, we have reduced the boundary problem (1.1)–(1.2) to the system of three algebraic equations for the four functions $v_{l\beta}(z_l)$ on the Riemann surface. Conversely, Eq. (1.11) implies that the distribution u_0, as defined by (1.8), is supported by $\overline{K}$ by the division theorem [52, 87]. Thus it remains to calculate all Cauchy data, i.e. four functions on the half-line from the three algebraic equations on the Riemann surface.

We eliminate two unknown functions from these three equations, which gives us one algebraic equation on the Riemann surface with two unknown functions. For example, we can eliminate the Dirichlet data v_{l0} and obtain the equation for the Neumann data

$$S_1(z)v_{11}(z_1) + S_2(z)v_{21}(z_2) = F(z), \qquad z \in V^*, \tag{1.12}$$

where the coefficients are the polynomials (12.7). This is an undetermined equation, so we should find some additional equations.

IV. Automorphic Functions on the Riemann Surface   We obtain the additional equations developing the method of automorphic functions, which was introduced by Malyshev in [77] in the context of difference equations in the quadrant of the plane. Additional equations are provided by the invariance of the functions $v_{l\beta}(z_l)$ with respect to the covering automorphisms of the Riemann surface.

We denote by $p_l : V \to \mathbb{C}$ with $l = 1, 2$ the coordinate projections $p_l : (z_1, z_2) \mapsto z_l$. Each projection p_l is a two-folded holomorphic *covering map*

$p_l : V \to \mathbb{C}$, and p_l^{-1} has two branching points, since the symbol $\tilde{A}(z)$ is irreducible by its strong ellipticity (7.6). Hence, the equation $z_l = p_l z$ with a fixed z_l admits two roots, $z', z'' \in V$, and these roots are distinct away from the branching points.

Definition 1.2 The monodromy automorphism $h_l : V \to V$ with $l = 1, 2$ is defined as

$$h_l(z') = z'' \qquad \text{and} \qquad h_l(z'') = z'$$

if $p_l z' = p_l z''$ and $z' \neq z''$. Otherwise, $h_l(z') = z'$. $\qquad\qquad\square$

In other words, the automorphism h_l transposes the points $z = (z_1, z_2)$ of the Riemann surface V with the same coordinates z_l. Now we can write new algebraic equations expressing the invariance of functions $v_{l\beta}(z)$ with respect to h_l:

$$v_{l\beta}^{h_l}(z) = v_{l\beta}(z), \quad z \in V_l^+, \qquad \beta = 0, 1, \quad l = 1, 2, \tag{1.13}$$

where $V_l^+ := \{z \in V : \operatorname{Im} z_l > 0\}$.

V. The Riemann-Hilbert Problem on the Riemann Surface Applying *formally* the automorphism h_1 to Eq. (1.12), we obtain

$$S_1^{h_1}(z)v_{11}(z_1) + S_2^{h_1}(z)v_{21}^{h_1}(z_2) = F^{h_1}(z) \tag{1.14}$$

since $v_{11}^{h_1} = v_{11}$ by (1.13). Now we can eliminate v_{11} from (1.12) and (1.14), and obtain the functional equation for one function v_{12},

$$Q_1(z)v_{21}(z) - Q_2(z)v_{21}^{h_1}(z) = G(z), \tag{1.15}$$

where $Q_1 = S_1^{h_1} S_2$, $Q_2 = S_1 S_2^{h_1}$, and $G = S_1^{h_1} F - S_1 F^{h_1}$. However, this equation is still an ill-posed problem due to some special topological peculiarities. The well-posed equation follows by using (1.13) with $l = 2$ for v_{21}. Now (1.15) becomes

$$Q_1(z)v_{21}(z) - Q_2(z)v_{21}^h(z) = G(z), \tag{1.16}$$

where $h := h_2 h_1$. This equation can be reduced to the Riemann-Hilbert problem in contrast to Eq. (14.8), which is the *one-sided Carleman problem* [150].

Remark 1.3 The justification of calculations (1.14)–(1.16) is nontrivial, since the region V^* in (1.12) is not invariant with respect to the automorphisms h_1 and h_2. $\square$

Further, we prove that the equation with the shift (1.16) is equivalent to the Riemann-Hilbert problem (16.6) on the factorspace V/h. To solve this problem we construct the factorization which is the solution to the corresponding homogeneous equation, and calculate the asymptotics of the factorization at infinity. This factorisation allows us to reduce the nonhomogeneous Riemann-Hilbert problem to the saltus problem (18.2) which is solved by the Cauchy type integral (18.9).

VI. The Reconstruction of Solutions to the Boundary Problem (1.1)–(1.2)
As the final step we reconstruct all solutions to the boundary problem (1.1)–(1.2) in the class of distributional solutions (7.22) and in the class of regular solutions (7.23), applying our calculations (1.14)–(1.16). As a consequence, we prove the Fredholmness of the boundary problem (1.1)–(1.2) in both classes of distributional and regular solutions.

VII. The Extension of the Method to Non-convex Angles with $\Phi > \pi$ This case differs significantly from the case $\Phi < \pi$. Namely, in the case $\Phi < \pi$ our method relies on the Paley-Wiener theory for analytic functions in the tube region (1.9) (see [54, 115, 122]) which provides a connection equation (1.11) on the complex characteristics. However, the corresponding tube region is empty in the case $\Phi > \pi$. Hence, we need some extra arguments to obtain a suitable version of the connection equation (1.11) in this case. In [58, 61] we developed the corresponding "dual arguments". Namely, we show that in the case $\Phi > \pi$ the same Connection Equation (1.11) holds *for the analytic continuation* of functions $v_{l\beta}$ *along the Riemann surface V*, see Chap. 20.

Remark 1.4 The MAF method gives a suitable extension of the Malyuzhinets approach from the Robin b.c. (6.1) to general b.c. (1.1), (1.2).   □

1.6   Development of the MAF Method

The methods of [52, 53] were developed in 1973–2007 by the authors for time-dependent diffraction problems. These problems are described by stationary boundary value problems of type (1.1)–(1.2) with the Helmholtz operator $A = \Delta + \omega^2$, where the frequency ω is real. However, the results of [52, 53] concern only the case of strongly elliptic operators corresponding to $\text{Im}\,\omega \neq 0$. The case of real ω required significant modifications. Namely, a direct application of the methods [52, 53] to the case $\omega \in \mathbb{R}$ meets two key difficulties:

1. The division theorem for elliptic operators [53] requires a suitable generalization to the operators $A = \Delta + \omega^2$ with $\omega \in \mathbb{R}$.
2. The corresponding Riemann-Hilbert problem (16.6) is ill-posed for $\omega \in \mathbb{R}$.

In [87], the required extension of the division theorem was established for arbitrary differential operators with constant coefficients. In [88], it was shown that the regularization $\omega \mapsto \omega \pm i0$ makes the Riemann-Hilbert problem well-posed. More precisely, the problem (1.1)–(1.2) with $A = \Delta + \omega^2$ and $\omega \in R$ is well-posed in the class of solutions $u(x, \omega)$ admitting a continuous extension to a solution $u(x, \omega \pm i\varepsilon)$ of the problem (1.1)–(1.2) with $A = \Delta + (\omega \pm i\varepsilon)^2$. This means the Limiting absorption principle for the problem (1.1)–(1.2) with $A = \Delta + \omega^2$.

1.7 Comments

The boundary problem in angles with general boundary conditions was proposed to AK by his supervisor Mark Vishik in the summer of 1968. AK introduced the MAF method during his studies in 1968–1972 at the chair of differential equations of the Faculty of Mechanics and Mathematics of Moscow State University. The results of his Ph.D. were published in [52, 53], and were reported in 1972 on seminars of M. I. Vishik, M. V. Fedoryuk and S. G. Mikhlin (who were the referees of the dissertation), and of V. A. Kondratiev and I. G. Petrovskii.

The research of AM started in 1971 at Mark Vishik's seminar at the Moscow State University. The problem of extension of the MAF method to the case of non - strongly elliptic Helmholtz operators was stated by AK in his talk at this seminar.

The results of AM's Ph.D. were published in [87, 88], and were reported in 1978–1980 on seminars of K. I. Babenko and M. V. Keldysh, and of M. V. Fedoryuk and V. P. Palamodov (who were the referees of the dissertation).

The subsequent collaboration of AK and AM resulted in novel development of the method in about 30 publications, which were discussed with J. L. Lions in 1980, with P. Grisvard in 1992 and with M. V. Babich in 2003.

1.8 Plan of the Book

Our book is organized as follows.

In Chap. 2 we outline the creation of the wave theory and diffraction of light by Grimaldi, Huygens, and Young. In Chap. 3 we present the Fresnel theory of diffraction and its justification and improvement by Kirchhoff. In particular, we present an update version of the Fresnel-Kirchhoff theory and its application to the diffraction by a half-plane.

In Chap. 4 we recall main principles of modern theory of stationary and time-dependent diffraction.

In Chap. 5 we present an update version of the Sommerfeld calculation of the diffraction by the half-plane [100, 129]. In Chap. 6 we give a concise survey of the results on stationary and time-dependent problems in angles that appeared after Sommerfeld's work.

In Chaps. 7–19 we present the MAF method in convex angles $\Phi < \pi$ for strongly elliptic second-order operators A. We extend the methods of [52, 53] to all solutions from the space of tempered distributions. All steps are described in detail: the complex Fourier transform, functional equation on the Riemann surface, Malyshev's method of automorphic functions, and the reduction to the Riemann-Hilbert problem.

Finally, in Chap. 20 we extend the MAF to the case of non-convex angles with $\Phi > \pi$.

In Appendix we consider the Sobolev spaces on the half-line.

Acknowledgements Since 1999 this research was supported by the Institute of Physics and Mathematics of the Michoacán University of St. Nicolas de Hidalgo in Morelia (México), since 2002 by the Faculty of Mathematics of Vienna University (Austria), and since 2007 by the Institute of Information Transmission Problems of RAS (Moscow, Russia).

The research of AK was partly supported by Austrian Science Fund (FWF): P28152-N35, and by Department of Mechanics and Mathematics of Moscow State University.

The research of AM was partly supported by Instituto de Física y Matemáticas de la Universidad Michoacana de S.Nicolás de Hidalgo (CIC), by the National Council of Science and Technology (CONACYT) and by PROMEP, México.

The authors are indebted to Peter Zhevandrov for useful discussions and remarks. The authors express their deep gratitude to Dr. José Eligio De la Paz Méndez for the great work on the design of the figures.

Part I
Survey of Diffraction Theory

Chapter 2
The Early Theory of Diffraction

In this chapter we trace the creation of the wave theory of light and of the theory of diffraction by Grimaldi, Huygens, Young, Fresnel, and Kirchhoff. We recall the calculation of the diffraction by half-plane in the Fresnel-Kirchhoff approximation [11, 32, 133].

However, we do not touch still the very important contributions by Young and Fresnel, which concern the polarization of light. Many interesting details can be found in [15, 29, 75, 144].

The first account of the diffraction is due to Leonardo da Vinci [74, p. 43], who observed a broadening of the shadow. The diffraction manifests itself as the formation of fringes of shadows. A similar phenomenon for water waves—the propagation of long waves behind obstacles—is well known from everyday life. The discovery of light diffraction and its analysis played a key role in universal acceptance of the wave theory of light.

2.1 The Grimaldi Observations

About 1660, Grimaldi discovered and observed systematically the interference and diffraction of light [43] and coined the word "diffraction", see [15, p. 96]:

Grimaldi completed writing the book [43] shortly before his death and it was published in 1665. The book begins with a description of Grimaldi's most famous discovery, namely the diffraction of light. He created a pinhole through which he allowed light from the sun to enter a darkened room and fall on a screen. The screen was at an angle so that the light produced an elliptical image on the screen. He placed a thin rod in the path of the light and measured the size of the shadow on the screen. He discovered that the shadow was larger than it should have been given conical nature of the beam. From this he argued that this effect was impossible if

© Springer Nature Switzerland AG 2019
A. Komech, A. Merzon, *Stationary Diffraction by Wedges*, Lecture Notes in Mathematics 2249, https://doi.org/10.1007/978-3-030-26699-8_2

light consisted of corpuscles so light must have a fluid form which bent round the object. He also noticed coloured bands near to the shadow of the rod. Each band had three components, a white broad central part with a narrow violet band on the side nearest the shadow with a narrow red band on its side a furthest from the shadow. He then described what effect was produced by placing obstacles of a different shape from the rod in the path of the cone of light. The name diffraction was chosen by Grimaldi because the effect reminded him of how a flowing fluid splits apart when a thin stick is placed in its path—the Latin diffraction means to *"break apart"*.

2.2   The Huygens Principle

In 1690, Huygens made a revolution in optics discovering the fundamental principle of light propagation, which explains basic optical phenomena (see [47]):

Every point on a wave-front may be considered as a source of secondary spherical wavelets, which spread out in the forward direction at the speed of light. The new wave-front is the tangential surface to all of these secondary wavelets.

Using this principle, Huygens explained for the first time the reflection law and also the refraction law, which were known since 984 from Persian scientist Ibn Sahl at the Baghdad court, and were rediscovered later by Thomas Harriot in 1602 and by Willebrord Snellius in 1621. These explanations were the great achievements of the Huygens theory.

2.3   The Young Theory of the Interference and Diffraction

Young was the first to explain the Newton rings and the colours of thin films by the interference of light (1801) and coined the word "interference", see [144]. He interpreted the colours of thin films as the result of interaction of light waves reflected from the top surface of the film with those which penetrate the film and are reflected from its back surface. This interpretation was inspired by Newton's explanation of anomalous flood-tides in the *Tonkin gulf* by the interaction of waves flowing through two straits of different lengths, see [147, p. 329] and [102, pp 240–242]. Young suggested similar interaction for waves of any nature. This universality manifests the concept of the *field* long before Faraday!

Using this interpretation of the film colours, Young calculated for the first time the wave-lengths of red and blue light by observation of Newton's rings: $\lambda = 0.65\,\mu$m for red light, $\lambda = 0.57\,\mu$m for yellow light, $\lambda = 0.51\,\mu$m for green light, and $\lambda = 0.47\,\mu$m for blue light, etc. This explanation and the measures were the great triumph of Young's theory.

Young explained for the first time the diffraction by the interference. He developed the Grimaldi experiment in a darkened room with a pinhole for sunlight, and observed the same coloured fringes near the shadow of a strip of paper and of

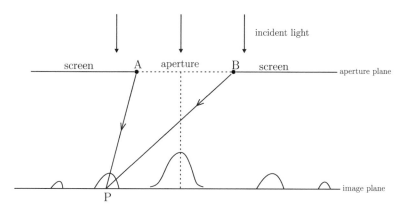

Fig. 2.1 Young diffraction

a horse hair [29]. His main novelty was the relocation of the strip to the boundary of the incident light cone: then the coloured fringes disappeared! The same result he observed when one of the sides of the strip was screened by a card. This fact allowed him to make his crucial discovery that the diffraction fringes result from the interference of secondary waves issued by different sides of the rod [29, 146, 147].

His general suggestion was that the diffraction results from the interference of secondary waves issued by the singularities of an obstacle. In particular, his calculation of diffraction on the slit relies on the interference of secondary waves issued by the edges A and B of the slit, see Fig. 2.1. The maxima and minima positions are given by the "Bragg rule" $|AP| - |BP| = n\lambda$ and $|AP| - |BP| = (n + 1/2)\lambda$ respectively with $n \in \mathbb{Z}$.

For the first time Young has performed the famous "two-slit experiment" observing the diffraction and the interference of the sunlight passing through two pinholes in a darkened room. This experiment definitely proved the wave nature of light.

Chapter 3
Fresnel–Kirchhoff Diffraction Theory

In this chapter we present the Fresnel theory of diffraction and its justification and improvement by Kirchhoff. In particular, we reproduce the calculation of the diffraction by the half-plane in the Fresnel-Kirchhoff theory.

3.1 Fresnel Theory of Diffraction

About 1814, Fresnel created the first complete mathematical theory of diffraction of light, which developed significantly the Young theory (though Fresnel until 1815 was not aware of Young's works). The Fresnel theory relies on the following two postulates.

I. The diffraction is due to the interference of secondary waves issued by all points of an aperture (and not only by the singularities of the obstacle, as in the Young theory).
II. The secondary wave issued by a point source P_0 with an amplitude A_0 reads

$$u(P, t) = \frac{A_0 e^{i(kr-\omega t)}}{r}, \tag{3.1}$$

where $r := |P - P_0|$, $k = \omega/c = 2\pi/\lambda$ is the wave number, and c is the light velocity.

The first postulate is a direct application of the Huygens principle. The second postulate is inspired by "harmonic plane waves" $u(x, t) = A_0 e^{i(kx_1-\omega t)}$ propagating in the direction x_1.

The factor $1/r$ requires special comments. Namely, the interference of light suggests that the "light intensity" $u(P, t)$ should take the positive and negative values. On the other hand, the luminance $E(P, t)$ is nonnegative, and "therefore" $E(P, t) \sim |u(P, t)|^2$. Another argument for this relation is the analogy with

© Springer Nature Switzerland AG 2019   19
A. Komech, A. Merzon, *Stationary Diffraction by Wedges*, Lecture Notes
in Mathematics 2249, https://doi.org/10.1007/978-3-030-26699-8_3

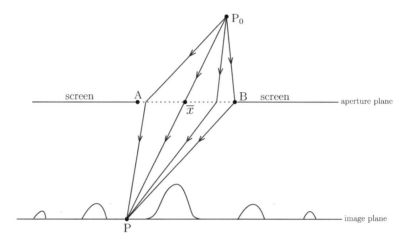

Fig. 3.1 The Fresnel diffraction

harmonic oscillator with potential energy $V = q^2$. Finally, it was well known empirically that $E(P, t) \sim r^{-2}$ for a point source, and hence, $|u(P, t)| \sim r^{-1}$.

The remarkable achievement of Fresnel was the calculation of diffraction on different apertures combining these postulates and developing ingenious mathematical technique. Namely, he considered the screen S which is a plate or an infinite plane, with some holes (apertures) occupying a region H of the plane. The light of frequency ω and of wavelength λ falls onto the screen from one side, while the point of observation lies on the other side of the screen, see Fig. 3.1.

The incident light is described by a harmonic incident wave, which produces the "light intensity" $A(x)e^{-i\omega t}$ at points $x \in H$ of the holes. Fresnel concludes from the postulates I and II that the light at the point P is described by the integral over the holes

$$u(P, t) = \int_H K(\chi) \frac{A(x)e^{i(ks-\omega t)}}{s} dS(x), \qquad (3.2)$$

where $s := |P - x|$, $\chi \in [0, \pi/2]$ denotes the angle between the vector $P - x$ and the incident ray at the point x, and $K(\chi)$ is an "inclination coefficient". Fresnel postulated that $|K(\chi)|$ decreases for $\chi \in [0, \pi/2]$, and $K(\pi/2) = 0$. Fresnel also suggests the value

$$K(0) = -\frac{i}{\lambda}, \qquad (3.3)$$

see Section 8.2 of [11]. However, to find an appropriate formula for $K(\chi)$ was a heavy problem in the framework of the Fresnel theory which was solved in the Kirchhoff theory, see next section.

Finally, in the case of the point source of frequency ω, located at the point P_0, the incident wave on the hole H is given by the postulate II:

$$A(x)e^{-i\omega t} = \frac{A_0 e^{i(kr-\omega t)}}{r},$$

where $r = |x - P_0|$ and A_0 denotes an intensity of the source. Substituting this into (3.2), we get

$$u(P,t) = A_0 e^{-i\omega t} \int_H K(\chi) \frac{e^{ik(r+s)}}{rs} dS(x) = A_0 e^{-i\omega t} \int_H K(\chi) \frac{e^{ik\Phi(x)}}{rs} dS(x).$$
$$(3.4)$$

where the phase function

$$\Phi(x) := r + s = |x - P_0| + |x - P|.$$
$$(3.5)$$

This is an oscillatory integral typical for the diffraction theory since Fresnel. The rectilinear path $P_0 \bar{x} P$ in Fig. 3.1 corresponds to the stationary point of the phase function, i.e.,

$$\nabla \Phi(\bar{x}) = 0.$$
$$(3.6)$$

Hence, the short-wave asymptotics $\sim k^{-1/2}$ for the integral (3.4) follows by the method of stationary phase [46, 115], when $\bar{x} \in H$. In this case the point of observation P is illuminated.

However, the high frequency asymptotics cannot explain the diffraction fringes, which are characteristic features of the diffraction of *rather long waves*. Fresnel invented special subtle method of *numerical integration* by writing

$$u(P,t) = A_0 e^{-i\omega t} e^{ik\Phi(\bar{x})} \int_H K(\chi) \frac{e^{ik[\Phi(x)-\Phi(\bar{x})]}}{rs} dS(x)$$

and partitioning the region of the integration H onto the "Fresnel Zones"

$$Z_n := \{x \in H : \Phi(x) - \Phi(\bar{x}) \in (n\pi, (n+1)\pi)\}, \qquad n = 0, 1, 2, \ldots, \qquad (3.7)$$

were the signs of the exponential $e^{ik[\Phi(x)-\Phi(\bar{x})]}$ are opposite for even and odd n. The zones are called "positive" for even $n = 0, 2, \ldots$ and "negative" for odd $n = 1, 3, \ldots$. Calculating the integral (3.4) over central zones with small numbers n, Fresnel approximates it by the oscillatory Fresnel integrals of type $\int e^{iakz^2} dz$, which arise due to (3.6). This Fresnel method of zones seems to anticipate the Lebesgue integration.

The partition (3.7) allowed Fresnel to calculate *numerically* the minimal and maximal points of the diffraction pattern for different apertures and the diffraction by thin wires, see [29, 41, 120]. In particular, the summation of the integrals (3.4) over the Fresnel Zones explained various paradoxical light phenomena [11, 32], for instance, the "Arago spot" (or Fresnel's bright spot) at the center of the shadow from a disk.

In 1816 Fresnel presented to the Paris Academy of Science a sketch of the manuscript, which was developed in the next 2 years into an exhaustive memoir [29, 41], submitted in 1818 for the Academy's prize.

He had calculated in the memoir the diffraction-patterns of a straight edge, of a narrow opaque body bounded by parallel sides and of a narrow opening bounded by parallel edges, and had shown that the results agreed excellently with his own experimental measurement (he measured with the micrometer the distance between the points with maximal and minimal luminance!).

Poisson was a member of the Committee and "... he studied Fresnel's theory in detail and, being a supporter of the Newtonian particle theory of light, looked for a way to prove it wrong. Poisson thought that he had found a flaw when he argued that a consequence of Fresnel's theory was that there would exist an on-axis bright spot in the shadow of a circular obstacle, where there should be complete darkness according to the particle theory of light. Since the Arago spot is not easily observed in everyday situations, Poisson interpreted it as an absurd result and that it should disprove Fresnel's theory.

However, the head of the Committee, Arago ... decided to perform the experiment in more detail. He molded a 2 mm metallic disk to a glass plate with wax. He succeeded in observing the predicted spot, which convinced most scientists of the wave nature of light and gave Fresnel the win" (https://en.wikipedia.org/wiki/Arago_spot).

As a result, Fresnel was awarded in 1818 the Grand Prix of Paris Academy of Science.

3.2 The Kirchhoff Theory of Diffraction

In 1883 Kirchhoff discovered the mathematical foundation of the Fresnel theory of diffraction and significantly improved this theory relying on the wave equation

$$\ddot{u}(x, t) = c^2 \Delta u(x, t), \qquad x \in \Omega := \mathbb{R}^3 \setminus S \tag{3.8}$$

as suggested by the Maxwell theory of light (1860). First, Kirchhoff obtained all the formulas (3.1), (3.2) and (3.4) directly from this equation. Moreover, in this way he identified the Fresnel coefficient $K(\chi)$ as

$$K = \frac{i}{2\lambda}[\cos \alpha_0(x) - \cos \alpha(x)], \qquad x \in H. \tag{3.9}$$

Here $\alpha_0(x) := (n(x), x - P_0)$ and $\alpha(x) := (n(x), x - P)$, where $n(x)$ denotes the unit normal to the aperture plane directed to the half-space containing the source point P_0. Fresnel mainly considered the case when $\alpha_0(x) \approx \pi$ and $\alpha(x) \approx 0$. Then also $\chi \approx 0$, so the Kirchhoff theory predicts the inclination coefficient

$$K(\chi) = -\frac{i}{2\lambda}[1 + \cos \chi], \qquad (3.10)$$

which perfectly agrees with (3.3) for small $|\chi|$. This prediction was a significant advantage of the Kirchhoff theory. Let $\Omega \subset \mathbb{R}^3$ denote the half space behind the aperture plane. For harmonic waves $u(x, t) = u(x)e^{-i\omega_0 t}$ the wave equation (3.8) becomes the stationary Helmholtz equation with the wave number $k = c/\omega_0$

$$(\Delta + k^2)u(x) = 0, \qquad x \in \Omega. \qquad (3.11)$$

The key Kirchhoff's observation was that the Fresnel wave of the point source (3.1) is proportional to the "elementary solution"

$$E(x) = -\frac{e^{ik|x|}}{4\pi |x|}$$

of the Helmholtz equation. In modern terms,

$$(\Delta + k^2)E(x) = \delta(x), \qquad x \in \mathbb{R}^3.$$

Hence, every solution to (3.11) can be expressed via its Cauchy data (Dirichlet and Neumann boundary values): by the Stokes theorem, for $P \in \Omega$,

$$u(P) = \int_{\partial\Omega} [\partial_n G(x, P)u(x) - G(x, P)\partial_n u(x)]dx \qquad (3.12)$$

where the Green function

$$G(x, P) := E(x - P) = -\frac{e^{ik|x-P|}}{4\pi |x - P|} \qquad (3.13)$$

and $\partial_n := n(y) \cdot \nabla_y$ is the differentiation along the external normal $n(y)$ to the boundary $\partial\Omega$ at the point $y \in \partial\Omega$. This formula follows by multiplying Eq. (3.8) by $G(P, x)$ and integrating by parts in the integral over the region behind the screen S where $|x| \leq R$. The integral over the large sphere with $|x| = R \to \infty$ cancels due to the Sommerfeld radiation conditions of type (4.22) for $u(x)$ and $G(x, P)$: for the three-dimensional case

$$(\partial_r - ik)u(x) = o(|x|^{-1}), \qquad |x| \to \infty.$$

Note that this cancellation holds for the "retarded Green function" (3.13), while it does not hold for the "advanced Green function" $\overline{G(x, P)}$, see the details in [5].

3.2.1   The Kirchhoff Approximation

To calculate the diffraction amplitude $u(x)$ from (3.12), we need to know the Cauchy data of the solution, i.e., $u(x)$ and $\partial_n u(x)$ for $x \in \partial\Omega$. The boundary of the region Ω consists of two components: (a) the holes H, which are illuminated by the point source located in front of the screen plane, and (b) the dark side of the screen $S = \partial\Omega \setminus H$, which is in the shadow.

The key idea of Kirchhoff was an *identification of the Cauchy data* on $\partial\Omega$. Namely, the experimental data and the Fresnel theory suggest that for large frequencies k.

1. the Cauchy data on the illuminated part H almost coincide with the corresponding data of the incident wave (3.1).
2. the Cauchy data on the dark part of the screen S vanish for large frequencies with high precision.

The error in both cases is concentrated near the border of the holes at the distances $\sim \lambda$. Thus the *Kirchhoff Cauchy data* read as follows:

$$
(u(x), \partial_n u(x)) \approx
\begin{cases}
A_0 e^{-ikt}(\frac{e^{ikr}}{r}, \partial_n \frac{e^{ikr}}{r}), & x \in H \\[2mm]
(0, 0), & x \in S
\end{cases}.
$$

Substituting these approximations into (3.12) we will obtain the approximate solution, which is very close to the exact solution at the points x with distance $\gg \lambda$ from the border of the holes. First, differentiation gives

$$
\partial_n \frac{e^{ikr}}{r} = \frac{e^{ikr}}{r}\left[ik + \frac{1}{r}\right]\cos(n(x), x - P_0)
$$

$$
= ik\frac{e^{ikr}}{r}\cos(n(x), x - P_0) + \mathcal{O}(r^{-2})
$$

Similarly,

$$
\partial_n G(x, P) = -\frac{ik}{4\pi}\frac{e^{iks}}{s}\cos(n(x), x - P) + \mathcal{O}(s^{-2}).
$$

Hence, substituting into (3.12), we obtain the *Fresnel-Kirchhoff diffraction formula*

$$
u(P) = \frac{ikA_0}{4\pi}\int_H \frac{e^{ik(r+s)}}{rs}[\cos\alpha_0(x) - \cos\alpha(x)]dx. \tag{3.14}
$$

This expression is proportional to the Fresnel integral (3.4) with the coefficient K given by (3.10) in the case $\alpha_0(x) \approx \pi$, $\alpha(x) \approx 0$, considered by Fresnel. The Fresnel theory and experiments suggest that this formula gives a good approximation if $\lambda \ll D := \text{diam}\,(H)$ and $\text{dist}\,(P, H) \gg \lambda$.

Thus, Kirchhoff's theory justified the Fresnel diffraction theory in the framework of stationary Helmholtz equation. Moreover, Kirchhoff's theory predicted the inclination coefficient (3.9) which is in accordance with the Fresnel formula (3.3).

Finally, the Kirchhoff theory corrected some problematic features of the Huygens-Fresnel theory, see [32, pp. 181–182]: in particular, the Huygens principle cannot explain why the secondary waves do not propagate in the backward direction.

3.2.2   The Fraunhofer Diffraction

The Fraunhofer diffraction corresponds to remote source and observer, when the distances $|x - P_0|$ and $|x - P|$ are large with respect to the size of the hole H:

$$|x - P_0|, \quad |x - P| \gg \text{diam}\,H. \tag{3.15}$$

In this case the integral (3.12) significantly simplifies

$$u(P) \sim \frac{k A_0}{4\pi} \frac{\cos\overline{\alpha} - \cos\overline{\alpha}_0}{\overline{r}\,\overline{s}} \int_H e^{ik(r+s)} dx, \tag{3.16}$$

where $\overline{\alpha} \approx (n(x), x - P)$, $\overline{\alpha}_0 \approx (n(x), x - P_0)$ and $\overline{r} \approx r, \overline{s} \approx s$ for $x \in H$.

In the case $\pi - \overline{\alpha}_0 \neq \overline{\alpha}$ we have $\overline{x} \notin H$, where $\overline{x}$ is the intersection of the interval $P_0 P$ with the image plane, see Fig. 3.1. Hence, the phase function $\Phi(x) = r + s$ is asymptotically linear in $x \in H$ for the case (3.15) since $\overline{x}$ is stationary point of the phase function (3.5). Finally, the integral (3.16) with the linear wave function reduces to the Fourier transform of the shape of the aperture H. Hence, it can be calculated easily for different holes H. For example, for the rectangle and for the circle, see Sections 8.5 and 8.6 of [11].

3.3   Diffraction by a Half-Plane in the Fresnel–Kirchhoff Theory

Fresnel calculated the diffraction by the half-plane screen

$$S := \{x \in \mathbb{R}^3 : x_3 = 0, \ x_1 > 0\}$$

calculating approximately the diffraction integral (3.4). In this case the aperture H is the complementary half-plane,

$$H = \{x \in \mathbb{R}^3 : x_3 = 0, \ x_1 < 0\}. \tag{3.17}$$

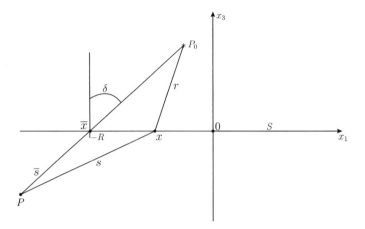

Fig. 3.2 The Fresnel diffraction by half-plane

We assume that the point source P_0 is located in the upper half-space $x_3 > 0$ and the point of observation P lies in the lower half-space $x_3 < 0$. Let R denote the distance of the stationary point $\bar{x}$ from the edge of the screen S, see Fig. 3.2.

Let us explain the Fresnel approximate method of calculation of the Fresnel-Kirchhoff integral (3.14). The calculations rely on the Taylor expansion of the phase function. By (3.6),

$$r + s = \Phi(x) \approx \Phi(\bar{x}) + \frac{1}{2}\Phi''(\bar{x})(\xi, \xi), \qquad x \in H, \tag{3.18}$$

where $\xi := x - \bar{x}$. This expansion reduces integral (3.14) to "Fresnel's integrals" of the type $\int e^{iakz^2}dz$.

Namely, let us split the integral (3.14) into the sum of two integrals: the first one over a small neighborhood $O(\bar{x})$ of the point $\bar{x}$ and the second one over the complement of this neighborhood. The second integral is negligible for large $|k|$ due to the high-frequency oscillations of the integrand. Hence,

$$u(P) \approx \frac{ikA_0}{4\pi} \int_{O(\bar{x})} \frac{e^{ik(r+s)}}{rs}[\cos\alpha_0(x) - \cos\alpha(x)]dx.$$

Substitute here the expansion (3.18), we obtain

$$u(P) \approx \frac{ikA_0}{4\pi}e^{i\Phi(\bar{x})} \int_{O(\bar{x})} \frac{e^{\frac{1}{2}ik\Phi''(\bar{x})(\xi,\xi)}}{rs}[\cos\alpha_0(x) - \cos\alpha(x)]dx. \tag{3.19}$$

Further, let us consider the case when the distances r and s are large compared to the diameter of the neighborhood $O(\overline{x})$. Then

$$\cos\alpha_0(x) \approx -\cos\delta \quad \text{and} \quad \cos\alpha_0(x) \approx \cos\delta \qquad \text{for} \quad x \in O(\overline{x}),$$

where δ is the angle between the line $P_0 P$ and the normal to the screen plane. Moreover,

$$r \approx \overline{r} := |P_0\overline{x}| \quad \text{and} \quad s \approx \overline{s} := |P\overline{x}| \qquad \text{for} \quad x \in O(\overline{x}).$$

Hence,

$$\cos\alpha_0(x) - \cos\alpha(x) \approx -2\cos\delta, \qquad \Phi(\overline{x}) \approx k(\overline{r}+\overline{s}).$$

Now (3.19) becomes the Fresnel-type integral

$$u(P) \approx -\frac{ikA_0\cos\delta}{2\pi\overline{r}\,\overline{s}}e^{ik(\overline{r}+\overline{s})}\int_{O(\overline{x})}e^{\frac{1}{2}ik\Phi''(\overline{x})(\xi,\xi)}dx.$$

Extending the integral over the aperture H, we obtain finally the Fresnel solution to the diffraction problem

$$u(P) \approx -iBe^{ik(\overline{r}+\overline{s})}\int_{H}e^{\frac{1}{2}ik\Phi''(\overline{x})(\xi,\xi)}dx, \qquad B := \frac{kA_0\cos\delta}{2\pi\overline{r}\,\overline{s}}, \tag{3.20}$$

since the integral over the complement of $O(\overline{x})$ is again negligible for large $|k|$ due to the high frequency oscillations.

Let us express solution (3.20) via the Fresnel integrals. First, calculate the Hessian $\Phi''(\overline{x})$ in the coordinates $\xi_1 := x_1 + R$, $\xi_2 := x_2$, with origin at the point $\overline{x}$. Differentiation gives

$$\Phi''(0)(\xi,\xi) = \left(\frac{1}{\overline{r}}+\frac{1}{\overline{s}}\right)[\xi_1^2\cos^2\delta + \xi_2^2], \qquad \xi = (\xi_1,\xi_2) \in \mathbb{R}^2.$$

In new coordinates the aperture (3.17) reads

$$H = \{\xi \in \mathbb{R}^2 : \xi_1 < R\},$$

and so (3.20) becomes

$$u(P) \approx -iBe^{ik(\overline{r}+\overline{s})}\int_{-\infty}^{R}\left[\int_{\mathbb{R}}e^{\frac{1}{2}ik(\frac{1}{\overline{r}}+\frac{1}{\overline{s}})(\xi_1^2\cos^2\delta+\xi_2^2)}d\xi_2\right]d\xi_1. \tag{3.21}$$

Let us change the variables

$$k\left(\frac{1}{r}+\frac{1}{s}\right)\xi_1^2\cos^2\delta = \pi\tau^2, \qquad k\left(\frac{1}{r}+\frac{1}{s}\right)\xi_2^2 = \pi\sigma^2, \qquad w = \sqrt{\frac{2}{\lambda}\left(\frac{1}{r}+\frac{1}{s}\right)}R\cos\delta.$$

$$\tag{3.22}$$

Then the integral (3.21) becomes

$$u(P) \approx -iBbe^{ik(\bar{r}+\bar{s})}\int_{-\infty}^{w}e^{i\frac{\pi}{2}\tau^2}d\tau\int_{\mathbb{R}}e^{i\frac{\pi}{2}\sigma^2}d\sigma, \qquad b := \frac{\pi}{k(\frac{1}{\bar{r}}+\frac{1}{\bar{s}})\cos\delta}.$$

$$\tag{3.23}$$

Here the last integral equals $1 + i$. Hence, calculating the product Bb we obtain the final *Fresnel-Kirchhoff formula*

$$u(P) \approx \frac{1-i}{2}\frac{A_0 e^{ik(\bar{r}+\bar{s})}}{(\bar{r}+\bar{s})}\int_{-\infty}^{w}e^{i\frac{\pi}{2}\tau^2}d\tau.$$

$$\tag{3.24}$$

This approximation, up to a factor, was obtained first by Fresnel in 1818 in the case $\delta = 0$, see [41, p. 176] (and p. 120 of English translation [29]).

For a fixed position P_0 of the source, the following asymptotics holds:

$$|u(P)| \sim \begin{cases} \dfrac{|A_0|}{\bar{r}+\bar{s}}, & R\to +\infty \\[2mm] 0, & R\to -\infty \end{cases}$$

since the Fresnel integral with $w = \infty$ equals $1 + i$. This agrees with the physical intuition, because the case $R \to +\infty$ corresponds to illuminated point P, while $R \to -\infty$ corresponds to a point P in the shadow. This transition from light to shadow is concentrated in a transitory zone of size $\sim \sqrt{\lambda}$ since $w \sim R/\sqrt{\lambda}$ by (3.22). This justifies the rectilinear light propagation for small wavelength λ.

In the transitory zone the function $u(P)$ is highly oscillating, and the maxima/minima of $|u(P)|$ correspond to the light/dark fringes of the diffraction pattern. Fresnel calculated positions of the fringes and compared then with his own experiments. The agreement was very good.

The oscillatory behaviour of the Fresnel integral is perfectly described by the Euler-Cornu spiral, see Fig. 3.3.

Namely, the spiral is the image of the map $\mathbb{R} \to \mathbb{C}$ defined by the Fresnel integral

$$w \mapsto F(w) := \int_0^w e^{i\frac{\pi}{2}\tau^2}d\tau.$$

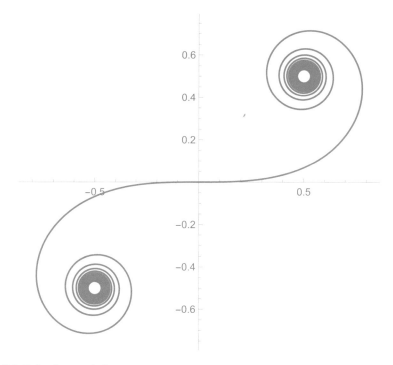

Fig. 3.3 Euler-Cornu spiral

The spiral converges to limits

$$F(w) \to F(\pm\infty) = \pm\frac{1+i}{2}, \qquad w \to \pm\infty. \qquad (3.25)$$

This spiral allows us to express the amplitude of the wave field (3.24) in the form

$$|u(P)| \approx \frac{|A_0|}{(\bar{r}+\bar{s})}|F(w) - F(-\infty)|/\sqrt{2}$$

which is highly effective for calculations.

It is important that the map F preserves the length: $|dF(w)| = |dw|$, since $dF = e^{i\frac{\pi}{2}w^2}dw$. Hence, the angular velocity (with respect to the length) of rotation of the point w on the spiral around the limit points $F(\pm\infty)$ is proportional to the radius $|F(w) \mp F(\pm\infty)|$. The rate of the convergence (3.25) is given by the asymptotics

$$F(w) - F(-\infty) = \int_{-\infty}^{w} e^{i\frac{\pi}{2}\tau^2}d\tau = \int_{w^2}^{\infty} e^{i\frac{\pi}{2}s}\frac{ds}{2\sqrt{s}}$$

$$= \int_{w^2}^{\infty} \frac{1}{\sqrt{s}}\frac{de^{i\frac{\pi}{2}s}}{i\pi} = \frac{e^{i\frac{\pi}{2}w^2}}{i\pi w} + \mathcal{O}(|w|^{-3/2}), \qquad w \to -\infty.$$

$$(3.26)$$

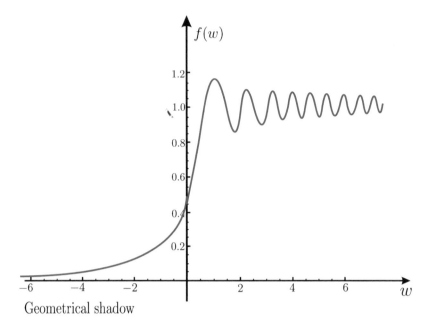

Fig. 3.4 Luminance near the light/shadow border

Finally, using this asymptotics and considering the spiral, it is easy to construct the graph of the amplitude $f(w) := |F(w) - F(-\infty)|/\sqrt{2}$ (see Fig. 3.4):

1. $f(w)$ vanishes as $w \to -\infty$.
2. $f(w)$ increases monotonically for $w \in (-\infty, 0)$.
3. $f(w)$ is highly oscillating for $w > 0$ and converges to the limit 1 as $w \to +\infty$. The asymptotics (3.26) implies that the period of these oscillations is proportional to their amplitudes, which are of the order $1/|w|$.

3.4 The Fraunhofer/Fresnel Limit: Diffraction of Plane Waves

The Fresnel-Kirchhoff approximation (3.23) is in a good agreement with experimental data, see [11] and [32]. Let us calculate the limit of (3.23) when the source point P_0 goes to infinity while the observation point P remains in the finite region. In this Fraunhofer/Fresnel limit the spherical incident wave (3.1) becomes a plane wave.

Hence, the limit of (3.23) gives a formula for the diffraction of plane waves by a half-plane. We will see in Sect. 5.8 that this limit of the Fresnel-Kirchhoff approximation (3.23) amazingly agrees with the exact formula obtained by Sommerfeld in 1896 [100, 129].

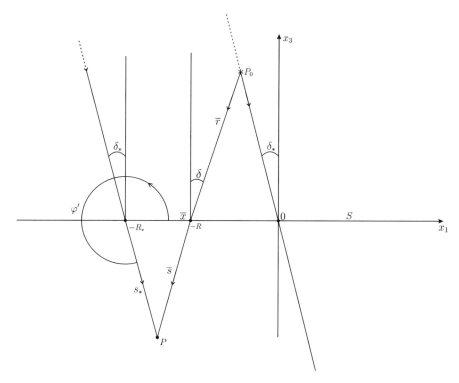

Fig. 3.5 The Fresnel-Fraunhofer diffraction by a half-plane

To calculate the Fraunhofer/Fresnel limit of diffraction let us move the source point P_0 to infinity in the plane $x_2 = 0$ along the ray $\varphi = \varphi_*$ with a fixed direction $\varphi_* = \pi/2 + \delta_* = \varphi' - 3\pi/2$. Thus, we take $P_0 = (r_* e^{i\varphi_*}, 0)$ with $r_* \to \infty$, see Fig. 3.5.

Now for the point of observation $P = (re^{i\varphi}, x_2) \in \mathbb{R}^3$ with $x_3 < 0$ we have

$$\delta \to \delta_*, \qquad \bar{r} + \bar{s} \sim r_* + r\cos(\varphi - \varphi'),$$

while

$$R \to R_* = -r(\sin\varphi \tan\delta_* + \cos\varphi), \qquad \bar{s} \to s_* := -r\sin\varphi/\cos\delta_*,$$

where $\varphi' = \varphi_* + \pi = 3\pi/2 + \delta_*$. We should keep the normalization

$$\frac{A_0(r_*)}{r_*} \sim a, \qquad r_* \to \infty \tag{3.27}$$

for $A_0 = A_0(r_*) \to \infty$ to obtain a nonzero limit for the luminance at the screen. In this limit

$$w = w(P_0, P) = \sqrt{\frac{2}{\lambda}\left(\frac{1}{r} + \frac{1}{s}\right)}R(P_0, P)\cos\delta \to w_* = \sqrt{\frac{2}{\lambda s_*}}R_*\cos\delta_* = \varkappa(\varphi)\sqrt{kr},$$

$$(3.28)$$

where

$$\varkappa(\varphi) := -\frac{(\sin\varphi\tan\delta_* + \cos\varphi)\cos\delta_*}{\sqrt{\pi(-\sin\varphi/\cos\delta_*)}}.$$

$$(3.29)$$

Remark 3.1 It is easy to check that

$$\begin{cases} \varkappa(\varphi) < 0, \ \varphi \in (\varphi', 2\pi) \\ \\ \varkappa(\varphi) > 0, \ \varphi \in (\pi, \varphi') \end{cases}.$$

$$(3.30)$$

□

Respectively, the diffraction amplitude (3.24) for $P = (re^{i\varphi}, x_2)$ reads asymptotically

$$u(P) \sim \frac{(1-i)a}{2}e^{ik(r_*+r\cos(\varphi-\varphi'))}\int_{-\infty}^{w_*}e^{i\frac{\pi}{2}\tau^2}d\tau, \qquad r_* \to \infty, \qquad (3.31)$$

which is independent of x_2.

Moreover, the amplitude of the spherical incident wave (3.1) becomes asymptotically the plane wave with the direction φ': (3.1) and (3.27) imply that for any fixed point $y = \rho e^{i\psi}$

$$\frac{A_0(r_*)e^{ik|y-P_0|}}{|y-P_0|} \sim \frac{A_0(r_*)e^{ik|y-P_0|}}{r_*} \sim ae^{ik(r_*+\rho\cos(\psi-\varphi'))}, \qquad r_* \to \infty, \qquad (3.32)$$

since

$$|y - P_0| = |\rho e^{i\psi} - r_* e^{i\varphi_*}| \sim r_* + \rho\cos(\psi - \varphi'), \qquad r_* \to \infty.$$

The diffraction amplitude (3.31) corresponds to the incident wave (3.32). Hence, eliminating the common factor e^{ikr_*} from (3.31) and (3.32) we obtain the formula

$$u(P) = \frac{(1-i)a}{2}e^{ikr\cos(\varphi-\varphi')}\int_{-\infty}^{w_*}e^{i\frac{\pi}{2}\tau^2}d\tau \qquad (3.33)$$

for the diffraction amplitude corresponding to the incident plane wave

$$u_i(y) = ae^{ik\rho\cos(\psi-\varphi')}.$$

(3.34)

3.5   Geometrical Optics and Diffraction

Formula (3.31) describes the rectilinear propagation of light and the broadening of the shadow. Namely, $re^{i\varphi} = -R_* + s_* e^{i\varphi'}$, and hence, (R_*, s_*) can be considered as the coordinates of the point of observation $P = x$ with $x_3 < 0$, and $|R_*|$ is proportional to the distance from the ray $\varphi = \varphi'$, which is geometrical light/shadow border, while s_* is proportional to $|x_3|$, see Fig. 3.6.

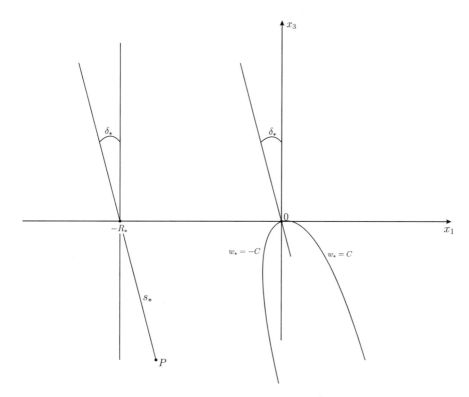

Fig. 3.6 The Fresnel-Fraunhofer diffraction by a half-plane

Formula (3.33) implies that the level lines of the luminance $|u(x)|$ coincide with those of w_*, which are semi-parabolas in the variables (R_*, s_*) due to (3.28):

$$\left|\begin{array}{ll} w_* = & C > 0 \text{ on the line} \quad \sqrt{\frac{2}{\lambda s_*}}R_* \cos \delta_* = \quad C \\ w_* = -C < 0 \text{ on the line} \quad \sqrt{\frac{2}{\lambda s_*}}R_* \cos \delta_* = -C \end{array}\right|. \tag{3.35}$$

Furthermore, (3.33) and (3.34) imply that

$$u(x) \sim \left\{\begin{array}{l} 0,\ w_* \to -\infty \\ u_i(x),\ w_* \to +\infty \end{array}\right|, \tag{3.36}$$

since the Fresnel integral (3.33) equals $1 + i$ for $w_* = \infty$.

Further, (3.29) implies that

$$\varkappa(\varphi) > 0 \quad \text{for} \quad \varphi \in (\pi, \varphi'), \qquad \varkappa(\varphi) < 0 \quad \text{for} \quad \varphi \in (\varphi', 2\pi). \tag{3.37}$$

Hence, using (3.28), this gives

$$\left\{\begin{array}{l} w_* \to -\infty,\ r \to \infty \text{ for} \quad \varphi \in (\varphi', 2\pi) \\ \\ w_* \to \infty, \quad r \to \infty \text{ for} \quad \varphi \in (\pi, \varphi') \end{array}\right|. \tag{3.38}$$

Now (3.36) implies that

$$u(x) \sim \left\{\begin{array}{ll} 0, & r \to \infty \text{ for} \quad \varphi \in (\varphi', 2\pi) \\ \\ u_i(x),\ r \to \infty \text{ for} & \varphi \in (\pi, \varphi') \end{array}\right|. \tag{3.39}$$

These asymptotics hold uniformly with a high precision outside the parabolic strip surrounded by the level lines (3.35) with large $C > 0$. Inside this parabolic strip the wave field goes from one limit value to another. This transition is oscillatory, and maxima and minima of $|u(x)|$ and perfectly described by the graph in Fig. 3.4. These maxima and minima satisfactory agree with the diffraction fridges which are observed experimentally, see [11, 32]. Furthermore, each parabolic strip shrinks to the ray (=half-plane) $\varphi = \varphi'$ as $\lambda \to 0$. The wave inside the strip becomes highly oscillating and converges to the discontinuous limit:

$$u(x) \to \left\{\begin{array}{ll} u_i(x),\ \varphi \in (\pi, \varphi') \\ 0,\ \varphi \in (\varphi', 2\pi) \end{array}\right|, \qquad \lambda \to 0. \tag{3.40}$$

Thus, for short wavelength λ the approximation (3.31) justifies the rectilinear light propagation that agrees with the geometrical optics. On the other hand, for each fixed $\lambda > 0$ the parabolic level lines (3.35) move away from the ray $\varphi = \varphi'$ as $s_* \to \infty$ which means the broadening of the shadow with the distance.

Finally, let us consider the diffracted wave, which is defined by

$$
u_d(x) := \begin{cases} u(x) - u_i(x), & \varphi \in (\pi, \varphi') \\ u(x), & \varphi \in (\varphi', 2\pi) \end{cases}. \tag{3.41}
$$

This diffracted wave is discontinuous on the ray $\varphi = \varphi'$ which is the light/shadow border. The jump is opposite to the jump of the incident wave, thus the total solution is continuous everywhere. Using (3.33) and (3.28), we find that

$$
u_d(re^{i\varphi}) = \begin{cases} \dfrac{(1-i)a}{2} e^{ikr\cos(\varphi-\varphi')} \displaystyle\int_{-\infty}^{\varkappa(\varphi)\sqrt{kr}} e^{i\frac{\pi}{2}\tau^2} d\tau, & \varphi \in (\varphi', 2\pi) \\[3mm] -\dfrac{(1-i)a}{2} e^{ikr\cos(\varphi-\varphi')} \displaystyle\int_{\varkappa(\varphi)\sqrt{kr}}^{\infty} e^{i\frac{\pi}{2}\tau^2} d\tau, & \varphi \in (\pi, \varphi') \end{cases}. \tag{3.42}
$$

Now the asymptotics (3.26) together with (3.30) imply the decay

$$
|u_d(re^{i\varphi})| \sim \frac{C}{|w_*|} \sim \frac{C}{|\varkappa(\varphi)|\sqrt{kr}}, \qquad r \to \infty \tag{3.43}
$$

Chapter 4
Stationary and Time-Dependent Diffraction

The *time-dependent diffraction* is aimed to determination of long-time asymptotics of wave processes, while the goal of *stationary diffraction* is the calculation of *limiting amplitudes*. Let us assume everywhere below in this book that the speed of light $c = 1$.

4.1 Time-Dependent Diffraction and Limiting Amplitude Principle

The *time-dependent diffraction* of harmonic plane waves by an obstacle $B \subset \mathbb{R}^3$ is described by the wave equation

$$\ddot{v}(x, t) = \Delta v(x, t), \qquad x \in \Omega := \mathbb{R}^3 \setminus B \tag{4.1}$$

with suitable boundary conditions on $\partial \Omega$, e.g., with the Dirichlet condition

$$v(x, t) = 0, \qquad x \in \partial \Omega, \tag{4.2}$$

and with the "initial condition"

$$v(x, t) = v_i(x, t), \qquad x \in \Omega, \quad t < 0. \tag{4.3}$$

For a *bounded* obstacle, v_i is an incident plane wave

$$v_i(x, t) := e^{-i(\omega_0 t - k_0 x)} F(\omega_0 t - k_0 x + R), \qquad x \in \Omega, \quad t < 0. \tag{4.4}$$

© Springer Nature Switzerland AG 2019
A. Komech, A. Merzon, *Stationary Diffraction by Wedges*, Lecture Notes
in Mathematics 2249, https://doi.org/10.1007/978-3-030-26699-8_4

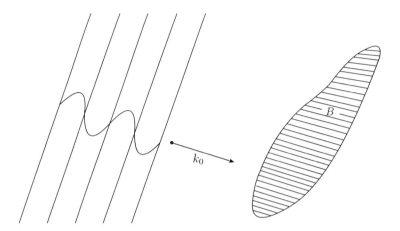

Fig. 4.1 The incident wave

Here $k_0 \in \mathbb{R}^3$ is the wave vector (Fig. 4.1) and $R = \min_{x \in B} k_0 x$, $\omega_0 = |k_0|$ is the frequency, and F stands for a profile function, i.e.,

$$F(s) = \begin{cases} 1, & s \geq \varepsilon \\ 0, & s \leq 0 \end{cases} \tag{4.5}$$

for an $\varepsilon > 0$. This profile function can be continuous or even a distribution. Then the incident wave (4.4) for $t < 0$ is the solution to (4.1) in the sense of distributions and satisfies the Dirichlet and Neumann boundary conditions on $\partial \Omega$.

Remark 4.1 For *unbounded* obstacles the initial data (4.4) should be completed by the corresponding *reflected waves*.   □

The *limiting amplitude principle* means the long-time asymptotics

$$v(x,t) \sim e^{-i\omega_0 t} u(x), \qquad t \to \infty, \quad x \in \Omega. \tag{4.6}$$

Remark 4.2 This asymptotics are well known for bounded obstacles B with sufficiently smooth boundary. In [35, 99] this asymptotics was established for convex and star-shaped obstacles B, and in [140]—for domains Ω satisfying the *non-trapping condition*.   □

Let us present general scheme of the proof of the asymptotics (4.6). First, similar asymptotics obviously holds for the incident wave (4.4): (4.5) implies that

$$v_i(x,t) \sim e^{-i\omega_0 t} u_i(x), \qquad x \in \Omega, \quad t \to \infty. \tag{4.7}$$

Next step is the construction of a suitable *reflected wave* with similar asymptotics

$$v_r(x, t) \sim e^{-i\omega_0 t} u_r(x), \qquad x \in \Omega, \quad t \to \infty. \tag{4.8}$$

The sum $v_i(x, t) + v_r(x, t)$ should satisfy needed boundary conditions (4.2). Hence, it remains to prove similar asymptotics for the *diffracted wave* $v_d(x, t) := v(x, t) - v_i(x, t) - v_r(x, t)$:

$$v_d(x, t) \sim e^{-i\omega_0 t} u_d(x), \qquad x \in \Omega, \quad t \to \infty. \tag{4.9}$$

The diffracted wave satisfies the wave equation

$$\ddot{v}_d(x, t) = \Delta v_d(x, t) + f(x, t), \qquad x \in \Omega := \mathbb{R}^3 \setminus B, \tag{4.10}$$

where $f(x, t) = (\Delta - \partial_t^2) v_r(x, t)$ since $(\Delta - \partial_t^2) v_i(x, t) \equiv 0$. Hence,

$$f(x, t) \sim f(x) e^{-i\omega_0 t}, \qquad f(x) = (\Delta + \omega_0^2) u_r(x) \tag{4.11}$$

due to (4.8). Moreover $v_d(x, t)$ satisfies the boundary condition (4.2).

Let us sketch the proof of the asymptotics (4.9) for solutions to Eq. (4.10) with $f(x, t) = f(x) e^{-i\omega_0 t}$. Thus, we consider equation

$$\ddot{v}_d(x, t) = \Delta v_d(x, t) + f(x) e^{-i\omega_0 t}, \qquad x \in \Omega. \tag{4.12}$$

We will assume $f \in L^2(\mathbb{R}^3)$ and zero initial data

$$v_d(x, 0) = \dot{v}_d(x, 0) = 0, \qquad x \in \Omega. \tag{4.13}$$

Let us apply the Fourier-Laplace transform in time defined as

$$\tilde{v}_d(x, \omega) = \int_0^\infty e^{i\omega t} v_d(x, t) dt, \qquad \text{Im } \omega > 0. \tag{4.14}$$

This integral converges, and it is analytic function in the complex upper half-plane as a function with the values in the Hilbert space $L^2(\mathbb{R}^3)$ since the L^2-norm of the solution v_d of (4.10) is bounded by $C(t + 1)$. This follows by standard arguments multiplying (4.10) by $\dot{v}_d(x, t)$ and integrating by parts, taking into account boundary condition (4.2). Applying the Fourier-Laplace transform to Eq. (4.12), we get *the stationary problem*

$$-\omega^2 \tilde{v}_d(x, \omega) = \Delta \tilde{v}_d(x, \omega) + \frac{f(x)}{i(\omega - \omega_0)}, \qquad x \in \Omega, \quad \text{Im } \omega > 0. \tag{4.15}$$

Therefore

$$\tilde{v}_d(\cdot, \omega) = \frac{R(\omega)f}{i(\omega - \omega_0)} = \frac{R(\omega_0 + i0)f}{i(\omega - \omega_0)} + \frac{R(\omega)f - R(\omega_0 + i0)f}{i(\omega - \omega_0)}, \qquad \text{Im } \omega > 0,$$

$$(4.16)$$

where $R(\omega) := (-\Delta - \omega^2)^{-1}$ is the resolvent corresponding to the Dirichlet boundary conditions (4.2). The last quotient in (4.16) is regular at $\omega = \omega_0$, and therefore its contribution is a dispersion wave. Hence, the asymptotics (4.9) holds with the limiting amplitude

$$u_d(x) = R(\omega_0 + i0)f. \qquad (4.17)$$

Remark 4.3 Thus, the time-dependent diffracted problem reduces to stationary b.v.p. in angles (4.15) with complex parameter ω. For time-dependent scattering by wedge the asymptotics (4.6) was established for the first time in [38, 60, 62, 66, 67]. The proofs follow general strategy above which reduces the time-dependent diffraction to the study of the corresponding stationary problem of the type (4.15) by the Fourier-Laplace transform (4.14). Our approach to this stationary problem relies on our general *method of automorphic functions* (MAF), presented in Chaps. 7–20. □

4.2   Stationary Diffraction Theory

The main goal of the *stationary diffraction theory* consists in the determination of the limiting amplitude $u(x)$. The limiting amplitude satisfies the stationary homogeneous Helmholtz equation

$$(\Delta + \omega_0^2)u(x) = 0, \qquad x \in \Omega, \qquad (4.18)$$

and the corresponding boundary conditions. Of course, these properties are insufficient for the determination of the limiting amplitude since we should take into account the initial condition (4.3) with the incident wave (4.4).

The first step towards the required characterization of the limiting amplitude is its splitting

$$u(x) = u_i(x) + u_r(x) + u_d(x), \qquad x \in \Omega, \qquad (4.19)$$

which follows from asymptotics (4.7)–(4.9). Then (4.18) reduces to the corresponding *inhomogeneous Helmholtz equation*

$$(\Delta + \omega_0^2)u_d(x) = -f(x), \qquad x \in \Omega, \qquad (4.20)$$

where $f(x) = (\Delta + \omega_0^2)u_r(x)$. This equation also admits an infinite set of solutions, so we need an additional selection rule to determine the required unique solution which is the limiting amplitude.

4.2.1 The Limiting Absorption Principle

The first selection rule which is known as the *limiting absorption principle* was introduced in 1905 by Ignatowsky [48]:

$$u_d(x) = \lim_{\varepsilon \to 0+} [\Delta + (\omega_0 + i\varepsilon)^2]^{-1} f(x), \qquad x \in \Omega; \qquad (4.21)$$

here the inverse operator is specified by the corresponding boundary conditionson $\partial\Omega$. The limit is known to exist in the case of sufficiently simple obstacles W (which are bounded and have a smooth boundary), and for equations with a potential, see [2, 35, 49, 57, 140]. Let us note that (4.21) coincides with (4.17).

4.2.2 The Sommerfeld Radiation Condition

An alternative (and equivalent under suitable conditions) selection rule is the *Sommerfeld radiation condition* introduced in 1912 by Sommerfeld [121, 131]: for 3D diffraction problems

$$(\partial_r - i\omega_0)u_d(r, \varphi) = o(r^{-1}), \qquad r \to \infty. \qquad (4.22)$$

Physically it means the absence of energy radiation from infinity. This asymptotics, together with the corresponding boundary conditions on $\partial\Omega$, provide the uniqueness of the solution to (4.20) if the obstacle B is bounded and the boundary Γ is sufficiently smooth. For 2D diffraction the corresponding radiation condition reads [4, Sect. 1.4]

$$\partial_r - i\omega_0 u_d(r, \varphi) = o(r^{-1/2}). \qquad (4.23)$$

The selection rules (4.21) and (4.22) are known to be equivalent to the limiting amplitude principle in many cases [35, 140].

For example, in the case of the wave equation (4.12) with a harmonic source the limiting amplitude satisfies both conditions (4.21) and (4.22): this easily follows from the Fourier-Laplace representation of the solution [136].

The radiation condition is very practical in numerical calculation of the limiting amplitude. On the other hand, the limiting absorption principle is more flexible for theoretical investigations. Equation (4.20), together with the corresponding boundary conditions and some of the selection rules, is called the *stationary diffraction problem* corresponding to the obstacle B.

Chapter 5
The Sommerfeld Theory of Diffraction by Half-Plane

In 1896 Sommerfeld calculated for the first time an exact solution to the stationary diffraction by the half-plane with the Dirichlet and Neumann b.c., see [100, 129]. His approach relies on:

1. The integral representation for solutions of the Helmholtz equation on the Riemann surfaces, and
2. The method of reflections on the Riemann surfaces.

Sommerfeld's approach to the problem earned him Poincaré's accolade: "méthode extrêmement ingénieuse [132, p. 675].

We present an updated version of Sommerfeld's calculations [11, 32, 133]. The Sommerfeld approach has not lost its relevance so far, since it gives the shortest way to the calculation of limiting amplitudes for the diffraction by wedges with rational angles $\Phi = \pi m/n$ for the Dirichlet and Neumann b.c. Our general MAF method gives the same result, but the way is longer. Moreover, our method is not applicable to the case $\Phi = 2\pi$ corresponding to the diffraction by the half-plane.

5.1 Stationary Diffraction by the Half-Plane

Sommerfeld considered the diffraction by a half-plane of the harmonic plane wave (4.4) with direction orthogonal to the edge of the half-plane. In this case, the diffraction is described by the 2D wave equation of type (4.1) in angle $Q = \mathbb{R}^2 \setminus \overline{S}$, where S denotes the screen $S := \mathbb{R}^+ \times 0$. Thus, Q is the angle of magnitude $\Phi = 2\pi$. Sommerfeld seeks the "standing wave solution" $v(x,t) \equiv u(x)e^{-ikt}$. Hence, the amplitude $u(x)$ is a solution to the stationary 2D Helmholtz equation

$$(\Delta + k^2)u(x) = 0, \qquad x \in Q. \tag{5.1}$$

Sommerfeld considered the Dirichlet and Neumann b.c.

© Springer Nature Switzerland AG 2019
A. Komech, A. Merzon, *Stationary Diffraction by Wedges*, Lecture Notes in Mathematics 2249, https://doi.org/10.1007/978-3-030-26699-8_5

Fig. 5.1 Diffraction by the
half-plane

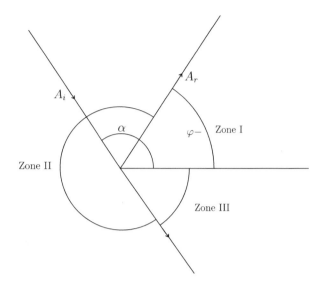

We will consider the diffraction by the half-plane for the incident wave (4.4) with
the directions

$$\varphi' \in (3\pi/2, 2\pi), \tag{5.2}$$

see Fig. 5.1. Sommerfeld introduces the "incidence angle" $\alpha := \varphi' - \pi$ instead of
φ'. According to (5.2),

$$\alpha \in (\pi/2, \pi). \tag{5.3}$$

The case $\alpha \in (\pi, 2\pi)$ reduces to (5.3) by the symmetry with respect to the plane of
the screen. In the cases $\alpha \in (0, \pi/2)$ and $\alpha \in (3\pi/2, 2\pi)$ the initial condition (4.3)
should be completed with the reflected waves (see Remark 4.1).

The case $\alpha = 0$ is considered in [38] for the Neumann b.c. The case $\alpha = \pi$ is
trivial for the Neumann b.c. and is rather singular for the Dirichlet b.c.

For concreteness, we will consider the Dirichlet b.c. In this case the corresponding amplitudes u_i and u_r of the incident and reflected waves in the polar coordinates
are given by geometrical optics:

$$u_i(r, \varphi) = \begin{cases} e^{ikr\cos(\varphi-\varphi')}, & \varphi \in (0, \varphi'), \\ 0, & \varphi \in (\varphi', 2\pi); \end{cases} \tag{5.4}$$

$$u_r(r, \varphi) = \begin{cases} -e^{ikr\cos(\varphi+\varphi')}, & \varphi \in (0, 2\pi - \varphi'), \\ 0, & \varphi \in (2\pi - \varphi', 2\pi) \end{cases} \tag{5.5}$$

where φ' is the direction of the incident wave. We will keep this notation till the end of this section.

5.2   Reflections on the Riemann Surface

The program of Sommerfeld was inspired by the well-known method of reflections, which gives solutions to boundary value problems in angles $\Phi = \pi/n$ with $n \in \mathbb{N}$. Namely, for $n = 1$ the solution in the half-plane is the difference of two solutions on the plane $\mathbb{R}^2$ corresponding to f and its reflection, for $n = 2$ the solution in the quarter plane is the alternating sum of the four reflected solutions on the plane $\mathbb{R}^2$ and so on. However, this method does not work for $\Phi \ne \pi/n$. In the diffraction by a half-plane we have $\Phi = 2\pi$, and in this case

> in ordinary space there is no place for the mirroring process. We therefore construct a double-covered space, in which the mirroring process is possible [100, p. 11].

Sommerfeld suggested to extend the "ordinary space" to the Riemann surface $\mathcal{R}$ of the function $\sqrt{z}$ and construct a "branching solution" by reflections on this Riemann surface, see the formula on page 319 of [129] (the formula on page 11 of [100]). Such a solution in the case $k = 0$ can be constructed by a conformal map, but the case $k \ne 0$ is much more difficult.

However, Sommerfeld does not follow this program. Instead, he constructs a universal integral representation for "branching solutions" to the homogeneous Helmholtz equation for the limiting amplitude (5.1),

$$(\Delta + k^2)u(x) = 0, \qquad x \in \mathcal{R}.$$

Afterward, he guessed in this integral representation a density providing the long range asymptotics corresponding to the incident and reflected waves (5.4), (5.5). Then he applies the method of reflections on the Riemann surface to provide the required boundary conditions, see (5.39) below.

5.3   Integral Representation for Branching Solutions

In 1892–1897, Poincaré analyzed polarization of light in diffraction using the separation of variables in polar coordinates for the Helmholtz equation, [113, 114]: the solution expands into series

$$u(r, \varphi) = \sum_{m}\{a_m \cos m\varphi + b_m \sin m\varphi\} J_m(kr), \qquad 0 < \varphi < 2\pi, \qquad (5.6)$$

where J_m is the Bessel function. Sommerfeld substituted the series (5.6) by a novel "plane-wave" integral representations for branching solutions to the Helmholtz equation. This integral representation was inspired (1) by the series (5.6) with non-integer m (which give the branching solutions), and (2) by the integral representation for the Bessel functions, as obtained by Hankel in 1869, see [142, Chapter VI]. The Sommerfeld construction relies on the Maxwell derivation of spherical functions [28], and on the Klein lectures on the Lamé functions 1889–1890.

5.3.1 The Maxwell Representation for Spherical Functions

Let us denote the plane $S := \{y \in \mathbb{R}^3 : y_3 = -1\}$. The crucial role in the Klein theory of spherical functions belongs to the *stereographic projection* $\mathbb{R}^3 \setminus S \to \mathbb{C}$ defined by formula (1) of [100, p. 22]:

$$y \mapsto \frac{y_1 + iy_2}{y_3 + |y|}, \qquad y = (y_1, y_2, y_3) \in \mathbb{R}^3 \setminus S. \qquad (5.7)$$

Restricted to the unit sphere $\Omega := \{\omega \in \mathbb{R}^3 : |\omega| = 1\}$, this projection is a conformal map $\Omega \setminus P_- \to \mathbb{C}$, where $P_- = (0, 0, -1)$ is the "South Pole". Hence, for any harmonic function $f(x)$ of $x \in \mathbb{C}$, the superposition

$$g(\omega) := f\left(\frac{\omega_1 + i\omega_2}{\omega_3 + 1}\right), \qquad \omega \in \Omega \setminus P_-$$

is the harmonic function on the unit sphere outside P_-. This means that

$$\Lambda g(\omega) = 0, \qquad \omega \in \Omega \setminus P_-.$$

where Λ is the *spherical Laplacian*, which is the spherical part of 3D Laplacian in the spherical coordinates:

$$\Delta_3 = \partial_r^2 + \frac{2}{r}\partial_r + \frac{1}{r^2}\Lambda. \qquad (5.8)$$

Let us extend $g(\omega)$ from the sphere Ω to $\mathbb{R}^3 \setminus S$ as zero order homogeneous function

$$\tilde{g}(y) := f\left(\frac{y_1 + iy_2}{y_3 + |y|}\right), \qquad y \in \mathbb{R}^3 \setminus S. \qquad (5.9)$$

Then (5.8) shows that $\tilde{g}(y)$ is a harmonic function on $\mathbb{R}^3 \setminus S$,

$$\Delta_3 \tilde{g}(y) = 0, \qquad \mathbb{R}^3 \setminus S. \qquad (5.10)$$

Next Sommerfeld considers the derivatives

$$\tilde{g}_m(y) := \partial_3^m \tilde{g}(y), \qquad m = 0, 1, 2, \ldots, \qquad y \in \mathbb{R}^3 \setminus S, \tag{5.11}$$

following Maxwell's introduction of the spherical functions in the Maxwell treatise [79] (Maxwell refers to Gauss and Laplace). These functions are obviously also harmonic in $\mathbb{R}^3 \setminus S$. This harmonicity easily implies that $g_m := \tilde{g}_m\big|_{\Omega}$ is the eigenfunction of Λ:

$$\Lambda g_m(\omega) = \lambda_m g_m(\omega), \qquad \omega \in \Omega \setminus S; \qquad \lambda_m = -m(m-1), \tag{5.12}$$

as was discovered by Maxwell [79]. Indeed, (5.10) in the spherical coordinates for $\tilde{g}_m(r\omega) = r^{-m} g_m(\omega)$ we obtain

$$0 = \Delta \tilde{g}_m(r\omega) = [(-m)(-m-1)r^{-m-2} + \frac{2}{r}(-m)r^{-m-1} + \frac{1}{r^2}\Lambda]g_m(\omega)$$

$$= r^{-m}[-\lambda_m + \Lambda]g_m(\omega). \tag{5.13}$$

5.3.2 The Sommerfeld Limit of Spherical Functions

We suggest a motivation for the Sommerfeld's limit process. Let us denote by $P_+ = (0, 0, 1)$ the North Pole of the sphere. Then (5.12) suggests that the restriction of $\tilde{g}_m$ onto the tangent plane $T_+ := \{y \in \mathbb{R}^3 : y_3 = 1\}$ is asymptotically the eigenfunction of 2D Laplace operator Δ_2 on this plane. Indeed, $\tilde{g}_m(x)$ is constant along the rays since it is zero order homogeneous function. In particular, $\partial_3^2 \tilde{g}_m(P_+ + x) = 0$ at the point $x = 0$. Hence,

$$\Delta_2 \tilde{g}_m(P_+ + x) \sim \lambda_m \tilde{g}_m(P_+ + x), \qquad x \to 0, \quad x \in E_0, \tag{5.14}$$

where $E_0 := \{y \in \mathbb{R}^3 : y_3 = 0\}$. Furthermore, Eq. (5.12) implies that $\tilde{g}_m(P_+ + x)$ is an oscillating function on E_0 with wavelength $\sim 1/m$. This suggests that for any fixed $k \in \mathbb{R}$ we could expect the convergence

$$C_m \tilde{g}_m(P_+ + \frac{kx}{m}) \to v(x), \qquad x \in E_0, \qquad m \to \infty, \tag{5.15}$$

with an appropriate normalizing coefficients C_m. Then

$$\Delta_2 v(x) = -k^2 v(x), \qquad x \in E_0. \tag{5.16}$$

Indeed, (5.14) implies that

$$[\Delta_2 \tilde{g}_m](P_+ + \frac{kx}{m}) \sim \lambda_m \tilde{g}_m (P_+ + \frac{kx}{m}), \qquad m \to \infty, \tag{5.17}$$

and hence,

$$\frac{m^2}{k^2} \Delta_2 [\tilde{g}_m (P_+ + \frac{kx}{m})] \sim \lambda_m \tilde{g}_m (P_+ + \frac{kx}{m}). \tag{5.18}$$

Finally, dividing by m^2 and using (5.15), we formally obtain in the limit $m \to \infty$ the Helmholtz equation (5.16) since $\lambda_m / m^2 \to -1$.

5.3.3   The Sommerfeld Integral Representation

In his paper [100, 129], Sommerfeld proved the convergence of type (5.15) for slightly different functions. First, he considered the function (5.9) for $y_1, y_2 \in \mathbb{R}$ and $y_3 \in \mathbb{C}$ with $|y| := \sqrt{y_1^2 + y_2^2 + y_3^2}$ assuming that f is an analytic function on $\mathbb{C}$. In this case the function (5.11) is analytic in $y_3 \in \mathbb{C}$ with two branching points $y_3 = \pm i \sqrt{y_1^2 + y_2^2}$. Hence, the function

$$\frac{f\left(\frac{y_1 + iy_2}{y_3 + |y|}\right) - f\left(\frac{y_1 + iy_2}{y_3 - |y|}\right)}{|y|} \tag{5.19}$$

is analytic in $y_3 \in \mathbb{C}$ and is single-valued, without the branching points. Finally, Sommerfeld introduces the function

$$\tilde{h}_m(y) := \frac{(-1)^{m+1}}{m!} \partial_3^m \frac{f\left(\frac{y_1 + iy_2}{y_3 + |y|}\right) - f\left(\frac{y_1 + iy_2}{y_3 - |y|}\right)}{|y|}, \tag{5.20}$$

which is also analytic in $y_3 \in \mathbb{C}$, and hence,

$$\tilde{h}_m(y) = \frac{1}{2\pi i} \int_C \left[f\left(\frac{y_1 + iy_2}{z + \rho}\right) - f\left(\frac{y_1 + iy_2}{z - \rho}\right) \right] \frac{1}{(y_3 - z)^{m+1}} \frac{dz}{\rho}, \tag{5.21}$$

where $\rho = \rho(y_1, y_2, z) := \sqrt{y_1^2 + y_2^2 + z^2}$ and C is any contour in $\mathbb{C}$ surrounding the point $y_3 \in \mathbb{C}$.

Therefore, the representation (5.21) holds for $y_3 = 1$ if C surrounds the point 1. In particular, (5.21) holds for

$$y = y(x, m) := P_+ + \frac{kx}{m} \tag{5.22}$$

with any fixed $x \in T_+$ if m is sufficiently large. Finally, Sommerfeld introduces the polar coordinates $x = re^{i\varphi}$ in T_+ and changes the integration variable

$$z = \frac{ikr}{m} \cos \gamma \in C, \tag{5.23}$$

where $\gamma \in \mathbb{C}$. In these coordinates

$$y(x, m) = (0, 0, 1) + \frac{kr}{m}(\cos \varphi, \sin \varphi, 0), \qquad y_1 + iy_2 = \frac{kr}{m}e^{i\varphi}, \qquad y_3 = 1$$

$$\rho = \sqrt{\frac{k^2 r^2}{m^2} + z^2} = \frac{kr}{m}\sqrt{1 - \cos^2 \gamma} = \frac{kr}{m} \sin \gamma.$$

Hence,

$$z \pm \rho = \frac{ikr}{m} \cos \gamma \pm \frac{kr}{m} \sin \gamma = \frac{kr}{m} e^{\mp i\gamma + \pi i/2}. \tag{5.24}$$

Substituting in (5.21), and taking into account that

$$\frac{1}{(y_3 - z)^{m+1}} = \frac{1}{(1 - \frac{ikr}{m} \cos \gamma)^{m+1}} \rightarrow e^{ikr \cos \gamma}, \qquad m \rightarrow \infty, \tag{5.25}$$

we obtain formally in this limit

$$\tilde{h}_m(y(x, m)) \rightarrow u(x) := \frac{1}{2\pi} \int_{C_\infty} \left[f(e^{i(\varphi - \gamma - \pi/2)}) - f(e^{i(\varphi + \gamma - \pi/2)}) \right] e^{ikr \cos \gamma} d\gamma, \quad m \rightarrow \infty, \tag{5.26}$$

where $x = re^{i\varphi}$ and C_∞ stands for the "limit contour". For example, such a limit contour exists when C is the circle centered at $\gamma = 1$ and of radius one. Then the limit contour reads $C_\infty = (i\infty, 0] \cup [0, \pi] \cup [\pi, i\infty)$. However, this choice of C is not optimal, since the resulting integral (5.26) generally diverges even for bounded functions f. Sommerfeld suggests to choose any contour C_∞ providing the convergence of the integral (5.26) for the function f under consideration.

Conclusion *The function*

$$u(r, \varphi) = \frac{1}{2\pi} \int_C f(e^{i(\varphi+\gamma)})e^{ikr\cos\gamma}d\gamma, \qquad r > 0, \quad \varphi \in [0, 2\pi] \qquad (5.27)$$

satisfies the Helmholtz equation

$$(\Delta + k^2)u(r, \varphi) = 0, \qquad r > 0, \quad \varphi \in [0, 2\pi] \qquad (5.28)$$

if the integral converges absolutely and locally uniformly in the parameters r and φ, as well as its first and second formal derivatives. In particular, f should be defined only on the contour C and does not need to be an analytic function.

5.4   Instructive Examples

5.4.1   Rational Density

As the first step, Sommerfeld considered the example

$$f(z) \equiv \frac{1}{1-z} \qquad (5.29)$$

[100, formula (1), p. 33]. Then integral (5.26) reads [100, formula (2), p. 34]

$$u(r, \varphi) = \frac{1}{2\pi} \int_C \left\{ \frac{1}{1 - e^{i(\varphi-\gamma)}} - \frac{1}{1 - e^{i(\varphi+\gamma)}} \right\} e^{ikr\cos\gamma}d\gamma$$

$$= -\frac{1}{2\pi i} \int_C \frac{\sin\gamma}{\cos\gamma - \cos\varphi} e^{ikr\cos\gamma}d\gamma. \qquad (5.30)$$

Here C denotes the contour

$$C = \mathcal{A} := (-\pi/2 + i\infty, -\pi/2 + i\varepsilon] \cup [-\pi/2 + i\varepsilon, 3\pi/2 + i\varepsilon] \cup [3\pi/2 + i\varepsilon, 3\pi/2 + i\infty) \qquad (5.31)$$

with a small $\varepsilon > 0$ and counter clock-wise direction, see Fig. 5.2. The integrals (5.30) converge since Im $\cos\gamma > 0$ for Re $\gamma \in ((2n-1)\pi, 2n\pi)$ if Im $\gamma > 0$. Changing the variables $\cos\gamma = a$ and $\cos\varphi = b$ we get

$$u(r, \varphi) = -\frac{1}{2\pi i} \int_{\mathcal{A}'} \frac{e^{ikra}da}{a - b}, \qquad (5.32)$$

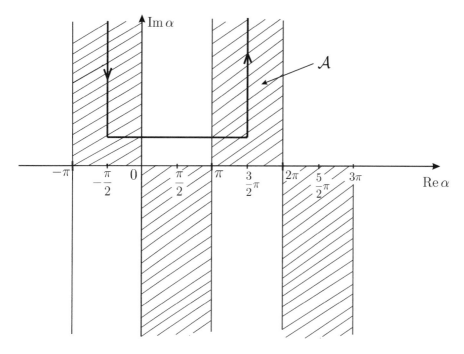

Fig. 5.2 Contour $\mathcal{A}$

where the contour $\mathcal{A}'$ surrounds the interval $[-1, 1]$. This is easily seen by calculating $\cos\gamma$ with $\gamma = -\pi/2 + i0, +i0, \pi/2 + i0, \pi + i0, 3\pi/2 + i0$. Hence, by the Cauchy theorem

$$u(r, \varphi) = e^{ikrb} = e^{ikr\cos\varphi} = e^{ikx_1}, \tag{5.33}$$

since the contour $\mathcal{A}'$ is oriented counter-clockwise. Now formula (5.30) shows that the function

$$f(z) \equiv \frac{1}{1 - z/z'} \tag{5.34}$$

with $z' = e^{i\varphi'}$ results in the plane wave with the direction φ':

$$u(r, \varphi) = e^{ikr\cos(\varphi - \varphi')}. \tag{5.35}$$

5.4.2   Branching Densities

As the next example, Sommerfeld considered the branching functions

$$f(z) = \frac{1/n}{1 - (z/z')^{1/n}} \tag{5.36}$$

with $n = 2, \ldots$ [100, formula (a), p. 35]. Taking the same contour (5.31), we obtain from (5.27),

$$u(r, \varphi - \varphi') = \frac{1}{2\pi} \int_{\mathcal{A}} \frac{e^{ikr \cos \gamma}}{1 - e^{i(\varphi - \varphi' + \gamma)/n}} d\gamma. \tag{5.37}$$

Here the integrand can be chosen to be a single-valued continuous function on $\mathcal{A}$. Nevertheless, the resulting function $u(r, \varphi - \varphi')$ is n-valued, since the integrand is not 2π-periodic in φ though it is $2\pi n$-periodic function in φ.

5.5   Diffraction by the Half-Plane

The examples above suggest the way to the calculation of the diffraction by a half-plane. First, the formula (5.35) suggests that the function $f(z)$ should have a pole at $z = z'$ with residue one, like (5.34). Second, the reflection strategy of Sect. 5.2 suggests that f should be branching two-valued function of type (5.37) with $n = 2$. Then the Sommerfeld integral with $f(z/z')$ and $z' = e^{\varphi'}$ should provide the desired branching solution on the double-sheeted Riemann surface $\mathcal{R}$.

First, Sommerfeld introduces the function

$$U(r, \psi) := \frac{1}{4\pi} \int_{\mathcal{C}} \frac{e^{i\frac{\gamma}{2}}}{e^{i\frac{\gamma}{2}} - e^{i\frac{\psi}{2}}} e^{-ikr \cos \gamma} d\gamma, \qquad \psi \in \mathbb{R} \bmod 4\pi, \tag{5.38}$$

where $\mathcal{C}$ is the Sommerfeld contour (two loops), see Fig. 5.3.

The function $e^{-ikr \cos \gamma}$ for $r > 0$ decreases super-exponentially on the contour $\mathcal{C}$ as $|\text{Im } \gamma| \longrightarrow +\infty$. Hence, the integral (5.38) converges, and it also converges after differentiation in r or φ any number of times.

Sommerfeld defines the solution to the diffraction problem by the antisymmetrization on the Riemann surface,

$$u(r, \varphi) := U(r, \varphi - \alpha) - U(r, \varphi + \alpha), \qquad \varphi \in \mathbb{R} \bmod 4\pi. \tag{5.39}$$

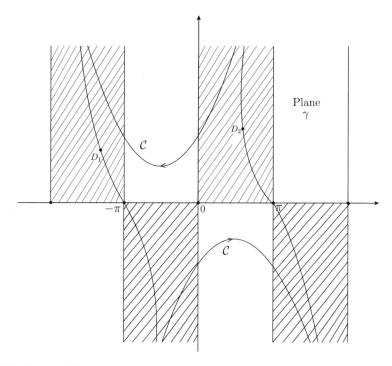

Fig. 5.3 Sommerfeld's contours

Now

$$
u(r, \varphi) = \frac{1}{4\pi} \int_C \left[\frac{e^{i\frac{\gamma}{2}}}{e^{i\frac{\gamma}{2}} - e^{i\frac{\varphi-\alpha}{2}}} - \frac{e^{i\frac{\gamma}{2}}}{e^{i\frac{\gamma}{2}} - e^{i\frac{\varphi+\alpha}{2}}} \right] e^{-ikr\cos\gamma} \, d\gamma
$$

$$
= \frac{1}{4\pi} \int_C \mathfrak{z}(\gamma, \varphi) e^{-ikr\cos\gamma} \, d\gamma, \quad \mathfrak{z}(\gamma, \varphi) := \frac{1}{1 - e^{i\frac{-\gamma+\varphi-\alpha}{2}}} - \frac{1}{1 - e^{i\frac{-\gamma+\varphi+\alpha}{2}}}.
$$

$$\tag{5.40}$$

Remark 5.1 Sommerfeld interprets this formula as the reflection on the Riemann surface [132, p. 252, formula (2)] which takes into account the incident and reflection waves (5.4) and (5.5).

Lemma 5.2 *The function $u(r, \varphi)$ satisfies the Helmholtz equation*

$$
(\Delta + k^2)u(r, \varphi) = 0, \quad r > 0, \ \varphi \in (0, 2\pi), \tag{5.41}
$$

and the Dirichlet boundary conditions

$$u(r, 0) = u(r, 2\pi) = 0, \qquad r > 0. \tag{5.42}$$

Proof

(i) Changing the variable of integration in (5.38), we obtain

$$U(r, \varphi - \alpha) = \frac{1}{4\pi} \int_{C_\varphi} \frac{e^{i\frac{\gamma}{2}}}{e^{i\frac{\gamma}{2}} - e^{-i\frac{\alpha}{2}}} e^{-ikr\cos(\gamma+\varphi)} \, d\gamma. \tag{5.43}$$

Here the contour C_φ can be fixed for small variations of φ by the Cauchy theorem. As a result, $(\Delta + k^2)U(r, \varphi - \alpha) = 0$ since

$$(\Delta + k^2)e^{-ikr\cos(\gamma+\varphi)} = 0. \tag{5.44}$$

Similarly, $(\Delta + k^2)U(r, \varphi + \alpha) = 0$.

(ii) (5.40) implies that

$$u(r, 0) = -\frac{i \sin \alpha/2}{4\pi} \int_C \frac{e^{-ikr\cos\gamma}}{\cos \gamma/2 - \cos \alpha/2} \, d\gamma = 0,$$

since the integrand is an even function and C is an "odd contour". The second boundary condition (5.42) follows similarly.   □

5.6   The Diffracted Wave

Here we calculate the diffracted wave u_d. We will (1) check the radiation condition (4.22) and the regularity condition (5.69), and (2) compare u_d with the Fresnel-Kirchhoff approximation (3.42).

First, let us transform the contour C in (5.40) into another contour $\Gamma = \Gamma_1 \cup \Gamma_2$, as shown in Fig. 5.4.

We should take into account the poles and the residues of the integrand. First, the poles of the integrand (5.38) are given by

$$\gamma_m = \psi + 4m\pi, \qquad m \in \mathbb{Z}, \tag{5.45}$$

and the corresponding residues are as follows

$$\text{Res}\left(\frac{e^{i\frac{\gamma}{2}}e^{-ikr\cos\gamma}}{e^{i\frac{\gamma}{2}} - e^{i\frac{\psi}{2}}}, \gamma_m\right) = -2ie^{-ikr\cos\psi}, \qquad m \in \mathbb{Z}. \tag{5.46}$$

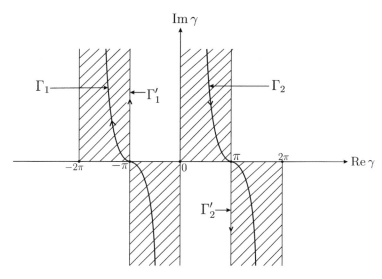

Fig. 5.4 The contour Γ

This implies that the poles of the integrand (5.40) are given by

$$\gamma_{m,1} = \varphi - \alpha + 4m\pi, \quad \gamma_{m,2} = \varphi + \alpha + 4m\pi, \quad m \in \mathbb{Z}, \tag{5.47}$$

with the corresponding residues

$$r_{m,1} = -2i e^{-ikr\cos(\varphi-\alpha)}, \quad r_{m,2} = 2i e^{-ikr\cos(\varphi+\alpha)}, \quad m \in \mathbb{Z}. \tag{5.48}$$

We partition the plane into the following zones (see Fig. 5.1):

I. $0 < \varphi < \pi - \alpha$,   **II.** $\pi - \alpha < \varphi < \pi + \alpha$,   **III.** $\pi + \alpha < \varphi < 2\pi$.

$$\tag{5.49}$$

In these notations the incident and reflected waves (5.4) and (5.5) read

$$u_i(r, \varphi) = \begin{cases} e^{-i|k|r\cos(\varphi-\alpha)}, & \varphi \in (0, \pi + \alpha), \\ 0, & \varphi \in (\pi + \alpha, 2\pi); \end{cases} \tag{5.50}$$

$$u_r(r, \varphi) = \begin{cases} -e^{-i|k|r\cos(\varphi+\alpha)}, & \varphi \in (0, \pi - \alpha), \\ 0, & \varphi \in (\pi - \alpha, 2\pi). \end{cases} \tag{5.51}$$

In [67] we have proved that the Sommerfeld solution (5.40) is the limiting amplitude for time-dependent diffraction. Hence, the following lemma implies that the diffracted wave (more precisely, its limiting amplitude) reads

$$u_d(r, \varphi) := u(r, \varphi) - u_i(r, \varphi) - u_r(r, \varphi) = \frac{1}{4\pi} \int_\Gamma \mathfrak{z}(\gamma, \varphi) e^{-ikr \cos\gamma} d\gamma, \quad \varphi \in (0, 2\pi) \backslash \{\pi - \alpha, \pi + \alpha\}. \tag{5.52}$$

Lemma 5.3 *The Sommerfeld solution (5.40) admits the following splitting in the zones I–III:*

$$\textbf{I.} \quad u(r, \varphi) = u_i(r, \varphi) + u_r(r, \varphi) + u_d(r, \varphi), \tag{5.53}$$

$$\textbf{II.} \quad u(r, \varphi) = u_i(r, \varphi) + u_d(r, \varphi), \tag{5.54}$$

$$\textbf{III.} \quad u(r, \varphi) = u_d(r, \varphi). \tag{5.55}$$

$\square$

Proof The formulas (5.47) and (5.3) imply that in zone **I** there are only two poles inside the contour Γ, namely,

$$\gamma_{0,1}, \gamma_{0,2} \in (-\pi, \pi). \tag{5.56}$$

Hence, (5.53) holds by (5.48) and the Cauchy residue theorem. In zone **II** only one pole $\gamma_{0,1} \in (-\pi, \pi)$ lies inside the contour Γ, which implies (5.54). Finally, in zone **III** none of the poles belongs to $(-\pi, \pi)$. Hence (5.55) follows.

Let us rewrite the diffracted wave u_d in a more convenient form. Let us denote

$$\mathfrak{z}_1(\gamma, \varphi) := \mathfrak{z}(\gamma - \pi, \varphi) - \mathfrak{z}(\gamma + \pi, \varphi)$$

$$= \frac{1}{1 - ie^{\frac{i(-\gamma + \varphi - \alpha)}{2}}} - \frac{1}{1 - ie^{\frac{i(-\gamma + \varphi + \alpha)}{2}}}$$

$$- \frac{1}{1 + ie^{\frac{i(-\gamma + \varphi - \alpha)}{2}}} + \frac{1}{1 + ie^{\frac{i(-\gamma + \varphi + \alpha)}{2}}}. \tag{5.57}$$

Lemma 5.4 *The diffracted wave u_d admits the representation*

$$u_d(r, \varphi) = \frac{i}{4\pi} \int_{\mathbb{R}} \mathfrak{z}_1(i\beta, \varphi) e^{ikr \cosh\beta} d\beta, \quad \varphi \in (0, 2\pi) \backslash \{\pi - \alpha, \pi + \alpha\}. \tag{5.58}$$

Proof Changing the variables of integration in (5.52), we obtain

$$u_d(r, \varphi) = \frac{1}{4\pi} \int_{\Gamma_1 + \pi} \mathfrak{z}_1(\gamma, \varphi) e^{ikr \cos\gamma} d\gamma, \quad \varphi \in (0, 2\pi) \backslash \{\pi - \alpha, \pi + \alpha\}. \tag{5.59}$$

Finally, we can transform the contour of integration $\Gamma_1 + \pi$ to $i\mathbb{R}$ by the Cauchy theorem since $|\mathfrak{Z}_1(\gamma, \varphi)| \sim e^{-|\gamma|/2}$ between these contours. □

The long-range asymptotics of the diffracted wave (5.58) follows by the method of stationary phase [26, (14.3) and (14.5))]:

$$u_d(r, \varphi) \sim \frac{i}{2}\sqrt{\frac{i}{2\pi}} \, \mathfrak{Z}_1(0, \varphi) \frac{e^{ikr}}{\sqrt{kr}}, \qquad r \to \infty, \quad \varphi \in (0, 2\pi) \setminus \{\pi - \alpha, \pi + \alpha\}.$$

$$(5.60)$$

5.7 Expression via Fresnel Integrals

As we will see in the next section, the total amplitudes (3.33) with $a = 1$ and (5.39) of the Sommerfeld solution and of the Fresnel-Kirchhoff approximation amazingly agree near the light/shadow border $\varphi = \varphi'$ for large r. It suffices to compare the diffraction amplitudes (5.58) and (3.42), since the incident waves (5.4) and (3.34) with $a = 1$ coincide.

To compare these diffraction amplitudes we will express the Sommerfeld diffracted wave (5.58) in the Fresnel integral of type (3.33) using the method of [11, 133]. First, we rewrite the function (5.57) as

$$\mathfrak{Z}_1(\gamma, \varphi) = S(\gamma, \varphi - \alpha) - S(\gamma, \varphi + \alpha),$$

where

$$S(\gamma, \psi) := \frac{1}{1 - ie^{i(-\gamma+\psi)/2}} - \frac{1}{1 + ie^{i(-\gamma+\psi)/2}} = \frac{e^{i\gamma/2}}{e^{i\gamma/2} - e^{i(\psi+\pi)/2}} - \frac{e^{i\gamma/2}}{e^{i\gamma/2} + e^{i(\psi+\pi)/2}}$$

$$= \frac{2e^{i(\gamma+\psi+\pi)/2}}{e^{i\gamma} - e^{i(\psi+\pi)}} = 2i\frac{e^{i(\gamma+\psi)/2}}{e^{i\gamma} + e^{i\psi}}.$$

Now the expression (5.52) becomes

$$u_d(r, \varphi) = \frac{1}{4\pi} \int_{\Gamma_1+\pi} [S(\gamma, \varphi - \alpha) - S(\gamma, \varphi + \alpha)]e^{ikr\cos\gamma} d\gamma, \qquad \varphi \in (0, 2\pi) \setminus \{\pi - \alpha, \pi + \alpha\}.$$

$$(5.61)$$

Let us denote

$$U_d(r, \psi) = \frac{1}{4\pi} \int_{\Gamma_1+\pi} S(\gamma, \psi)e^{ikr\cos\gamma} d\gamma, \qquad \psi \in (-\pi, 3\pi) \setminus \{\pi\}. \qquad (5.62)$$

Then similarly to (5.60),

$$|U_d(r, \psi)| \sim \frac{C(\psi)}{\sqrt{r}}, \qquad r \to \infty, \qquad \psi \in (-\pi, 3\pi) \setminus \{\pi\}. \tag{5.63}$$

Using the symmetry of the contour $\Gamma_1 + \pi$ with respect to $\gamma = 0$, we can rewrite (5.62) as the integral over the upper branch $\Gamma_1^+ + \pi$ of the contour $\Gamma_1 + \pi$,

$$U_d(r, \psi) = \frac{1}{4\pi} \int_{\Gamma_1^+ + \pi} [S(\gamma, \psi) + S(-\gamma, \psi)] e^{ikr \cos \gamma} d\gamma = \frac{i \cos \frac{\psi}{2}}{\pi} \int_{\Gamma_1^+ + \pi} \frac{\cos \frac{\gamma}{2}}{\cos \psi + \cos \gamma} e^{ikr \cos \gamma} d\gamma.$$

Multiplying by $e^{ikr \cos \psi}$ and differentiating in r, we get

$$\partial_r [e^{ikr \cos \psi} U_d(r, \psi)] = -\frac{k \cos \frac{\psi}{2}}{\pi} \int_{\Gamma_1^+ + \pi} \cos \frac{\gamma}{2} e^{ikr[\cos \psi + \cos \gamma]} d\gamma.$$

Finally, changing the variables $\sin \gamma/2 = \tau$, we obtain

$$\partial_r [e^{ikr \cos \psi} U_d(r, \psi)] = -\frac{2k \cos \frac{\psi}{2}}{\pi} e^{ikr \cos \psi} \int_{\tilde{\Gamma}} e^{ikr[1 - 2\tau^2]} d\tau.$$

Now dragging the contour $\tilde{\Gamma}$ to $i\mathbb{R}^+$ by the Cauchy theorem and changing afterwards $\tau = i\beta$ with $\beta \in \mathbb{R}^+$, we obtain

$$\partial_r [e^{ikr \cos \psi} U_d(r, \psi)] = -\frac{2ik \cos \frac{\psi}{2}}{\pi} e^{ikr[\cos \psi + 1]} \int_0^\infty e^{2ikr\beta^2} d\beta$$

$$= \frac{1 - i}{2} \sqrt{\frac{k}{\pi r}} \cos \frac{\psi}{2} e^{2ikr \cos^2 \frac{\psi}{2}} = \frac{1 - i}{2} \partial_r \int_{-\infty}^\rho e^{i\pi\tau^2/2} d\tau, \quad \rho = 2\sqrt{\frac{kr}{\pi}} \cos \frac{\psi}{2} =: v(\psi)\sqrt{r},$$

$$\tag{5.64}$$

since the first integral is $\sqrt{\pi/kr}/2 \times (1 + i)/2$. Hence,

$$U_d(r, \psi) = \frac{1 - i}{2} e^{-ikr \cos \psi} \left[\int_{-\infty}^\rho e^{i\pi\tau^2/2} d\tau + C(\psi) \right]. \tag{5.65}$$

5.8 Agreement with the Fresnel-Kirchhoff Approximation

Let us check that the Fresnel-Kirchhoff approximation (3.33) agrees with the Sommerfeld solution (5.39) for large $r > 0$. The incident and the reflected waves coincide. So, it suffices to compare only the diffracted waves. Their agreement is obvious for the directions $\varphi \neq \varphi'$ since there both diffracted waves decay like $1/\sqrt{r}$ by (5.63) and (3.43). It remains to consider the directions with $|\varphi - \varphi'| < \varepsilon$ for a small $\varepsilon > 0$. Using the notation (5.62), the diffracted wave (5.61) can be written as

$$u_d(r, \varphi) = U_d(r, \varphi - \alpha) - U_d(r, \varphi + \alpha).$$

We will prove the agreement of (5.61) with (3.42) for $\varphi > \varphi'$ and $\varphi < \varphi'$ separately.

(I) First let us consider the case $\varphi' < \varphi < \varphi' + \varepsilon$. The constant $C(\psi)$ in (5.65) is given by

$$C(\psi) = 0, \qquad \psi \in (\pi, 3\pi). \qquad (5.66)$$

Indeed, for $\psi \in (\pi, 3\pi)$ we have $v(\psi) < 0$, and hence, $\rho \to -\infty$ as $r \to \infty$. Now (5.66) follows from (5.62). Furthermore, $\alpha \in (\pi/2, \pi)$ by (5.3), and hence (5.66) implies that

$$U_d(r, \varphi \pm \alpha) = \frac{1-i}{2} e^{-ikr \cos(\varphi \pm \alpha)} \int_{-\infty}^{v_\pm(\varphi)\sqrt{kr}} e^{i\pi\tau^2/2} d\tau, \qquad \varphi \in (\varphi', 2\pi),$$

where

$$v_\pm(\varphi) = \frac{2}{\sqrt{\pi}} \cos \frac{\varphi \pm \alpha}{2}.$$

For $\varphi = \varphi' := \pi + \alpha$ we have

$$v_-(\varphi') = \frac{2}{\sqrt{\pi}} \cos \frac{\pi}{2} = 0, \qquad v_+(\varphi') = \frac{2}{\sqrt{\pi}} \cos \frac{\pi + 2\alpha}{2} = -\frac{2}{\sqrt{\pi}} \sin \alpha < 0.$$

Hence, $|U_d(r, \varphi + \alpha)|$ decays like $1/\sqrt{kr}$ in a neighborhood of $\varphi = \varphi'$ by (3.26). On the other hand, $U_d(r, \varphi - \alpha)$ decays slower in a neighborhood of $\varphi = \varphi'$, and does not decay at all for $\varphi = \varphi'$. As a result, we can neglect $U_d(r, \varphi + \alpha)$ for large r and $0 < \varphi - \varphi' \leq \varepsilon$ if $\varepsilon > 0$ is sufficiently small:

$$u_d(r, \varphi) \approx U_d(r, \varphi - \alpha) = \frac{1-i}{2} e^{-ikr \cos(\varphi - \alpha)} \int_{-\infty}^{v_-(\varphi)\sqrt{kr}} e^{i\pi\tau^2/2} d\tau, \quad 0 < \varphi - \varphi' \leq \varepsilon, \quad r \gg 1.$$

Finally, this expression almost coincides with the first line of (3.42) with $a = 1$ by the following lemma.

Lemma 5.5

(i) *For $\alpha \in (0, \pi)$,*

$$v_-(\varphi) - \varkappa(\varphi) = \mathcal{O}(|\varphi - \varphi'|^2), \qquad \varphi \to \varphi'. \qquad (5.67)$$

(ii) *In the case $\alpha = \pi/2$*

$$v_-(\varphi) - \varkappa(\varphi) = \mathcal{O}(|\varphi - \varphi'|^3), \qquad \varphi \to \varphi'. \tag{5.68}$$

Proof Let us denote $\varepsilon := \varphi - \varphi'$. Then $\varphi - \alpha = \pi + \varepsilon$, and

$$v_-(\varphi) = -\frac{2\sin \varepsilon/2}{\sqrt{\pi}}.$$

Let us calculate $\varkappa(\varphi)$ in terms of α and ε. First, (3.29) can be written as

$$\varkappa(\varphi) = \frac{\sin(\varphi - \varphi')(\cos \delta_*)^{3/2}}{\sqrt{\pi}\sin(\varphi')\sqrt{-\sin \varphi}}.$$

Here $\varphi' = \pi + \alpha$ and $\varphi = \varphi' + \varepsilon$, and hence,

$$\varkappa(\varphi) = \frac{\sin \varepsilon \cos(\alpha - \pi/2)\sqrt{\cos(\alpha - \pi/2)}}{\sqrt{\pi}\sin(\alpha + \pi)\sqrt{-\sin(\varepsilon + \pi + \alpha)}} = -\frac{\sin \varepsilon \sqrt{\sin \alpha}}{\sqrt{\pi}\sqrt{\sin(\alpha + \varepsilon)}}.$$

Now

$$v_-(\varphi) - \varkappa(\varphi) = -\frac{2}{\sqrt{\pi}}\sin \varepsilon/2\left[1 - \cos \varepsilon/2\sqrt{\frac{\sin \alpha}{\sin(\alpha + \varepsilon)}}\right],$$

which implies (5.67) and (5.68).   □

(II) In the case $\varphi' - \varepsilon < \varphi < \varphi'$ with a small $\varepsilon > 0$ we have $\varphi - \alpha \in (0, \pi)$. Now $v(\psi) > 0$ for $\psi \in (0, \pi)$ in notation (5.64), and hence, $\rho \to \infty$ as $r \to \infty$. In this case (5.65) and (5.63) give

$$U_d(r, \varphi - \alpha) = -\frac{1 - i}{2}e^{-ikr\cos(\varphi - \alpha)}\int_{v_-(\varphi)}^{\infty} e^{i\pi\tau^2/2}d\tau, \qquad \varphi' - \varepsilon < \varphi < \varphi'$$

However, (5.68) implies that this expression almost coincides with the second line of (3.42) with $a = 1$ by Lemma 5.5. Finally, the term $U_d(r, \varphi + \alpha)$ is negligible as before in a neighborhood of $\varphi = \varphi'$ for large $r > 0$.

Remark 5.6 The difference between the Fresnel-Kirchhoff approximation and the exact Sommerfeld solution in the zone $|\varphi - \varphi'| \le \varepsilon$ decays for small $\varepsilon > 0$ by Lemma 5.5. Moreover, the value of ε can be enlarged for larger $\sin \alpha$: the maximal ε corresponds to $\alpha = \pi/2$. Fresnel calculated and measured experimentally the diffraction by the half-plane exactly for the case $\alpha = \pi/2$.

5.9   Selection Rules for the Sommerfeld Solution

The Sommerfeld solution (5.40) appears up to this point as a particular solution to the homogeneous Helmholtz equation (5.41) satisfying the Dirichlet b.c. However, its relation to the time-dependent diffraction of the incident wave (4.4) was an open problem for about 100 years.

In the paper [100, 129] Sommerfeld proved the decay

$$u_d(r, \varphi) \to 0, \qquad r \to \infty, \qquad \varphi \in (0, 2\pi)$$

for the "diffracted" wave $u_d(r, \varphi) := u(r, \varphi) - u_i(r, \varphi) - u_r(r, \varphi)$ which is given by (5.52) according to Lemma 5.3. However, this decay alone is insufficient for the identification of the Sommerfeld solution $u(r, \varphi)$ with the limiting amplitude.

In the book [133] Sommerfeld checked both radiation conditions(4.23) and the Meixner regularity condition on the edge

$$r|\nabla u_d(x)| \to 0, \qquad r \to 0, \quad x \in Q \tag{5.69}$$

for the "diffracted wave" $u_d := u - u_i - u_r$ given by (5.59). The radiation condition (4.23) *for the asymptotics* (5.60) follows by direct differentiation, since $\omega = |k|$. To prove (4.23) for the *exact diffracted wave* $u_d(r, \varphi)$ it suffices to differentiate the integral representation (5.59) and apply afterwards the method of stationary phase. Finally, differentiating the integral representation (5.65), we obtain the estimate

$$\partial_r u_d(r, \varphi) = \mathcal{O}(1/\sqrt{r}), \qquad r \to 0, \tag{5.70}$$

which implies the regularity condition (5.69). The radiation condition (4.23) and regularity condition (5.69) provide the uniqueness of the solution to Helmholtz equation (4.20), see [4]. Hence, the Sommerfeld solution should coincide with the limiting amplitude if the limiting amplitude principle (4.6) would hold.

In [67] we have proved for the first time that the Sommerfeld solution is *the limiting amplitude* for the time-dependent diffraction by the half-plane of harmonic plane waves (4.4) with a broad class of profile functions.

Chapter 6
Diffraction by Wedge After Sommerfeld's Article

We give a concise survey of main results on diffraction by wedges and on related problems in angles after the Sommerfeld work [129].

6.1 Stationary Problems in Angles

In 1901, Sommerfeld extended his method to stationary diffraction by wedges of any rational magnitude $\Phi = \pi m/n$ with the Dirichlet and Neumann b.c., see [130, p. 38]. In 1920, Carslaw extended this Sommerfeld result to arbitrary angles $\Phi \in (0, 2\pi)$, see [16].

In 1954 Oberhettinger constructed the Green functions for the Helmholtz equation in the plane angle of any magnitude $\Phi \in (0, 2\pi)$ with the DD and NN boundary conditions, see [104]. In [105] he constructed the asymptotics of these Green functions for small and large distances from the edge, and near the light/shadow border. These results justify the asymptotics obtained previously by Pauli [109].

The stationary theory of diffraction by wedge for Maxwell field is presented in monographs [5, 11, 25, 133].

In 1958, Malyuzhinets obtained for the first time the formula for solutions to the stationary diffraction problem in angles with the impedance (Robin) b.c.

$$\partial_n u(x) + i a_l u(x) = 0, \qquad x \in \Gamma_l, \quad l = 1, 2 \tag{6.1}$$

with $a_l \in \mathbb{R}$, see [76, 78]. An updated version of the Malyuzhinets theory can be found in [4]. The problem is reduced first to a functional-difference equations with a shift, and then to the Riemann–Hilbert problem on a cylinder, which is solved in quadratures. The constructed solution satisfies the Sommerfeld radiation condition at infinity and the Meixner regularity condition (5.69) at the edge of the wedge.

© Springer Nature Switzerland AG 2019   63
A. Komech, A. Merzon, *Stationary Diffraction by Wedges*, Lecture Notes
in Mathematics 2249, https://doi.org/10.1007/978-3-030-26699-8_6

The resulting formula is represented in the form of a *Sommerfeld integral*. These *Sommerfeld–Malyuzhinets representations* were used in many applied problems (see, e.g., [6, 7]).

In 1958, Sobolev considered mixed problems in plane angles with the Cauchy data on one side and the Dirichlet data on another side. He formulated the criterion of well-posedness in terms of the corresponding eigenvalues [128]. In 1961, Shilov developed this result formulating appropriate conditions on the boundary values, [123].

The completeness of Ursell's trapped modes in angles was studied since 1950 by Peters [111], Roseau [116], Lehman and Lewy [73] and others.

In 1967–1970, Kondrat'ev obtained very important asymptotics of solutions to elliptic equations near the wedges of the domains with a piecewise smooth boundary [68, 69].

In 1970, Malyshev discovered a new powerful method of automorphic functions for solving the boundary value problem for difference equation in the quadrant of the plane which is a discrete approximation of Eq. (4.18), see [77]. This method uses complex Fourier transform in two variables on the lattice and Galois theory in the ring of analytic functions on the corresponding Riemann surface. As the result, the boundary value problem reduces to a functional equation with a shift, which in turn is reduced to the Riemann–Hilbert problem using the Galois automorphisms.

In 1971, Maz'ya and Plamenevskii [80] calculated exact solution to the oblique derivative b.c. for the equation $(\Delta - \lambda)u(x) = f(x)$ with $\lambda > 0$ in angles of arbitrary magnitude $\Phi \in (0, 2\pi)$ The method uses the Fourier transform in $r = |x|$ and reduces the problem to a Riemann–Hilbert problem with discontinuous b.c., which is solved in quadratures.

In 1973, one of the authors introduced a novel method of "automorphic functions on complex characteristics" [52, 53]. This method gives exact solutions for general second-order elliptic equations (1.1) in convex angle $Q \subset \mathbb{R}^2$ of magnitude $\Phi < \pi$, with general boundary conditions (1.2). In 1992, the method was extended by the authors to non-convex angle with $\Phi > \pi$, see [58, 61].

This method was applied in 1977–2018 to various problems:

1. Time-dependent scattering by wedges with the Dirichlet and Neumann b.c. , see next section.
2. The problem of the completeness of Urcell's trapped modes on a sloping beach [63, 89].
3. The existence and stability of Ursell's trapped modes for two-layer fluid [148].
4. The existence and uniqueness of solution to the Helmholtz equation (4.18) with Im $\omega \neq 0$ in angles of any magnitude $\Phi \in (0, 2\pi)$ under the Neumann b.c. [149]. Previously this problem was solved only for the right angle [84].
5. The Fredholmness of boundary value problems with wedges [52].

In 1975, Maz'ya and Plamenevskii extended their method [80] to the equations $(\Delta - \lambda)u(x) = f(x)$ for $\lambda > 0$ in any angle $\Phi \in (0, 2\pi)$ with general differential boundary conditions with real coefficients, [81]. The method [81] uses essentially the fact that all coefficient are real and $\lambda > 0$. Thus, this method is non-applicable to the calculation of the diffracted wave (4.21) where $\lambda = -(\omega + i\varepsilon)^2$.

In a series of papers 1974–1987 and in the monograph [44], Grisvard studied second-order elliptic boundary value problems for Laplacian in curvilinear polygons $\Omega \subset \mathbb{R}^2$, with the boundary $\partial\Omega = \cup_k \Gamma_k$, where Γ_k are diffeomorphic to the interval $[0, 1]$,

$$
\begin{cases}
\Delta u(x) = f(x), & x \in \Omega \\
u\big|_{\Gamma_k} = 0, & k \in \mathcal{D}; \quad [\partial_{n_k} u + w_k \partial_{\tau_k}]\big|_{\Gamma_k} = 0, \quad k \in \mathcal{N}
\end{cases}
\tag{6.2}
$$

Here n_k and τ_k denote the unit normal tangent vectors to Γ_k, and $w_k \in \mathbb{R}$. Solutions are considered from the Sobolev spaces $W_p^2(\Omega)$. The complete description of traces of the function from this space is given in [44, Sect 1.5]. One of central results is Theorem 4.4.1.2 of [44] which implies in particular that

1. The problem has at most one solution in $W_p^2(\Omega)$ defined up to an additive constant, and
2. The problem has at most one solution in $W_p^2(\Omega)$ if $\mathcal{D} \neq \emptyset$.

The existence and uniqueness of solutions from the space $H^1(\Omega)$ is proved for every $f \in L_p(\Omega)$ if $\mathcal{D} \neq \emptyset$. For the case $\int_\Omega f \, dx = 0$ the solution in the space $H^1(\Omega)$ exists if $\mathcal{D} = \emptyset$, and it is unique up to an additive constant (Lemma 4.4.3.1 of [44]). This result allows one to calculate the index of the problem (6.2) in appropriate spaces using Kondrat'ev's technique [68, 69], see Theorem 4.4.37 of [44].

In 1985, Costabel and Stephan proved asymptotic error estimates for the finite element Galerkin approximation of the boundary integral equations corresponding to the Laplacian in a plane polygonal domain with the mixed Dirichlet–Neumann b.c., [27]. The approach relies on the Mellin transform and the Goarding inequality generalising the results [143] of Wendland, Stephan and Hsiao which concern the case of domains with smooth boundary.

In 1988, Dauge developed in [30] a general theory of boundary value problems in regions with corners generalising the results of Grisvard, Kondrat'ev, Maz'ya and others.

In 1987–2006, Bernard applied the Sommerfeld integral representation to asymptotic expansion and numerical simulation for diffraction by wedges and polygons with various types of boundary conditions , [6, 7].

The book [101] deals with general elliptic boundary value problems in domains with edges of different dimensions. It includes a survey of results of Kondrat'ev, Maz'ya, Nazarov and Plamenevskii on the Fredholmness in weighted Sobolev spaces and asymptotic expansions of solutions near singularities of the boundary.

In 1993, Bonnet-Ben Dhia and Joly considered stationary problem for three-dimensional linear water waves guided along the coast, [9]. The problem is reduced to a two-dimensional eigenvalue problem in an angle domain for a family of unbounded selfadjoint operators with noncompact resolvent. The existence and properties of the guided modes are considered with particular attention to low- and high-frequency behaviour. The authors point out a large variety of results that greatly depend on the geometry of the bottom relief.

In 1999, Bonnet-Bendhia, Dauge, and Ramdani studied spectral properties of non-coercive transmission problem arising in superconducting waveguides with angular interface, [10]. They considered the problem of existence of selfadjoint extensions, which are of crucial importance for numerical calculations of the corresponding eigenvalues. The criterion for the existence is given in terms of dielectric constant. The approach relies on the calculation of von Neumann indices using the Kondrat'ev asymptotics [68].

In 1987–2014, Meister, Speck, and their collaborators developed a novel approach to boundary value problems for the Helmholtz equation (4.18) with $\omega \in \mathbb{C} \setminus \mathbb{R}$ in the quadrant $Q = \mathbb{R}^+ \times \mathbb{R}^+$ for the Dirichlet and Neumann b.c., see [18, 19, 22, 42, 82, 84, 86]. The solution u is considered in the Sobolev space $H^s(Q)$ with $s \in [1, 3/2)$. The boundary data $f := (f_1, f_2)$ belong to space $H^{s_1}(\partial Q)$, which is the trace space of $H^s(Q)$. The description of this space is based on Grisvard's results [44]. For example, $s_1 = 1/2$ for $s = 1$, and $H^{1/2}(\partial Q) = \{(f_1, f_2) \in H^{1/2}(\Gamma_1) \times H^{1/2}(\Gamma_2) : f_1 - f_2 \in \tilde{H}^{1/2}(\mathbb{R}^+)\}$, where the definition of $\tilde{H}^{1/2}$ can be found, for example in [149]. The results include the existence, uniqueness, well-posedness and an explicit representation of solutions.

Similar results were obtained for solutions $u \in H^1(Q)$ to the "transmission problems" in the angles of the magnitudes $\Phi = \pi/2$, and $\Phi = 3\pi/2$, [23, 82–85]. Later these results were extended to some "rational" angles: for the angles $\Phi = \pi/4$ in [18], for $\Phi = 3\pi/2$ in [20], for $\Phi = 4\pi/3$ in [103], for $\Phi = 2\pi$ in [33]. In [34] these results were extended to any rational angles $\Phi = n\pi/m$, using the Riemann surface approach.

In 1999, Penzel and Teixeira extended these results to the Robin b.c. (6.1) for $\Phi = \pi/2$ in the Sobolev space H^1, see [110]. The uniqueness was proved and the explicit formula for solution was given if a certain matrix admits a canonical factorization. In [17, 19] the existence and uniqueness of solutions for $\Phi = \pi/2$ were proved. However, this problem with the Robin b.c. is open till now for general angles $\Phi \in (0, 2\pi)$.

In 2014, Castro and Kapanadze [21] extended these results to boundary value problems in arbitrary angles $\Phi \in (0, 2\pi]$ with the Dirichlet and Neumann b.c. The existence, uniqueness and well-posedness were proved, and explicit integral representations for solutions was written in terms of a solution to a 1D pseudodifferential equation.

6.2 Time-Dependent Diffraction by Wedges

The nonstationary diffraction by wedges was much less studied.

In 1901, Sommerfeld obtained in [130] the first formula for time-dependent diffraction by a wedge with the rational angles $\Phi = \pi m/n$ for incident plane wave (4.4) with $\omega = 0$ and general profile function F. The limiting amplitude principle was not discussed.

In 1932–1937, Smirnov and Sobolev [125–127] obtained a formula for the diffraction by a wedge of the incident wave (4.4) with $\omega = 0$ and the Heaviside profile function $F(s) = \theta(s)$: in this case the incident wave reads

$$v_i(x, t) = \theta(t - k \cdot x). \tag{6.3}$$

They considered wedges of arbitrary magnitude $\Phi \in (0, 2\pi]$ with the Dirichlet and Neumann b.c. The derivation relies on the Smirnov–Sobolev representation of solutions to the wave equation by analytic functions, see [124].

In 1951, Keller and Blank considered the same problem developing Busemann's "Conical Flow Method" which is similar to Smirnov–Sobolev's approach, see [51]. In 1953, Kay developed an alternative approach to similar problem for general plane wave using separation of variables, see [50]. The solution was represented in the form of the Whittaker functions series [145, p. 279]. The author proved that the series coincide with Keller-Blank solution from [51] in the case of the incident wave (6.3), see [50, p. 434].

In 1958, Oberhettinger considered time-dependent diffraction by wedges with the Dirichlet and Neumann b.c. and general incident plane waves (4.4) with $\omega \neq 0$, [106]. The solution to the corresponding stationary problem is constructed in integral form using modified Hankel functions [106, (11)]. The final formula for the time-dependent problem is a convolution with the corresponding kernel, see [106, (108)]. The existence and uniqueness of solutions in appropriate function classes were not considered.

In 1958, Petrashen', Kouzov and Nikolaev established for the first time the limiting amplitude principle for a particular solution corresponding to the incident wave (4.4) for an arbitrary angle $\Phi \in (0, 2\pi)$ with the Dirichlet and Neumann b.c., see [112]. An explicit solution of this problem was obtained by separation of variables and the solution is given as a series. Its coincidence with Sobolev's solution for incident wave (6.3) was claimed. However, the existence and uniqueness of the solution from a functional class was not discussed.

In 1961, Papadopoulos extended in [108] the Sobolev–Keller–Blank approach calculating the nonstationary diffraction of the pulse (6.3) by an imperfectly reflecting wedge with the Leontovich boundary condition [72]

$$\partial_n u(x) = a \partial_t u(x), \qquad x \in \Gamma. \tag{6.4}$$

In 1966, Borovikov constructed the Green functions for time-dependent scattering by wedge with the Dirichlet and Neumann b.c., assuming the Sommerfeld-type representations [12]. He constructed a particular solution for incident wave (4.4) with $F(s) = s_+^{-1/2}$, and reproduced the Sobolev's solution [125].

In 1967, Filippov considered the time-dependent scattering problem by wedge with the impedance boundary conditions (6.4), [40]. The author got rid of restrictions on the wedge angle and the direction of incident wave from previous paper [119]. Moreover the author constructed the ray expansion of the solution near the wave front, calculating all terms of high-frequency asymptotics of the corresponding stationary problem (the first term is found in [76]; the entire expansion was obtained in [105], but under certain restrictions). The convergence of the ray expansion is improved by separation of singularities. Moreover, the author studied the solution near the light/shadow border, and the surface waves in the non-stationary problem (the stationary case has been studied in [76]). The uniqueness of the solution was proved specifying its behavior near the edge of the wedge.

In 1992, Eskin considered the wave equation in an n-dimensional dihedral angle of arbitrary magnitude $0 < \Phi < 2\pi$ with pseudodifferential b.c., see [36]. The Fourier transform in time and along the edge of the angle reduces the problem to a system of type (1.1), (1.2) in a plane angle. Further steps are parallel to our approach of 1973 (see [52] and Sect. 1.5)

In 1993, Gérard and Lebeau applied the microlocal analysis technique to the diffraction by a curvilinear wedge under the Dirichlet b.c., see [42]. The diffracted wave was represented in the form of an oscillatory integral for small $|x| + |t|$. An asymptotic series for the amplitude of this integral was calculated. The main result is a description of the propagation of singularities for small $|x| + |t|$.

In 1998, Rottbrand considered the diffraction of plane wave (4.4) by the wedge with DN b.c. The problem was reduced by a conformal map to the Rawlins's mixed problem, and the solution was represented by an infinite series of Bessel functions, see [117, Sect. 3] and [118].

In 2002, Lafitte extended the methods and results [42] to the impedance b.c. [71].

In 1996–2018, the MAF method was applied by the authors together with Zhevandrov and others to time-dependent scattering of incident plane waves by wedges with the Dirichlet and Neumann b.c. The main results are as follows:

1. The existence, uniqueness and explicit formulas for solutions in diffraction by wedges in appropriate functional classes [31, 37, 38, 60, 65, 92].
2. The *limiting amplitude principle* in diffraction by wedges and formulas for the corresponding *limiting amplitudes*, [31, 38, 60, 66, 92].
3. Justification of the formulas obtained by Sobolev, Keller, Blank and Kay [50, 51, 124, 125]. Namely, we proved in [62, 97] the coincidence of these solutions with the corresponding limiting amplitudes. This coincidence implies the stability of these particular solutions under perturbation of the profile of incident plane wave.
4. The long-time asymptotics of the diffracted wave [24] and the asymptotics near the wave front [24, 98].

In 2011, Hewett, Ockendon, and Allwright established the rate of convergence to the limiting amplitude in the diffraction by wedges with Neumann b.c. [45]. These results were extended in [24, 98] to other boundary conditions.

Part II
Method of Automorphic Functions
on Complex Characteristics

Chapter 7
Stationary Boundary Value Problems in Convex Angles

In the remaining part of the book we present a universal method of automorphic functions on complex characteristics. This method gives in particular all solutions from the space of tempered distributions to stationary problems of the type (4.15) in angles $Q \subset \mathbb{R}^2$. In Chaps. 7–19 we consider convex angles of magnitude $\Phi < \pi$, and in Chap. 20 we give necessary modifications for nonconvex angles with $\Phi > \pi$.

Thus, we consider the equation of type (4.15) in an angle Q of magnitude $\Phi < \pi$. We should solve the nonhomogeneous equation of type (4.15) with homogeneous boundary conditions. Subtracting a particular solution to the equation, we reduce the problem to solution of the homogeneous equation with nonhomogeneous boundary conditions. For simplicity of notation, we transform the angle Q into the first quadrant $K := \mathbb{R}^+ \times \mathbb{R}^+$ by a linear change of variables. Finally, we will consider equations

$$A(\partial)u(x) = \sum_{|\alpha| \leq 2} a^\alpha \partial^\alpha u(x) = 0, \qquad x \in K \tag{7.1}$$

with general boundary conditions

$$\begin{cases} B_1(\partial)u(x_1, 0+) = \sum_{|\alpha| \leq m_1} b_1^\alpha \partial^\alpha u(x_1, 0+) = f_1(x_1), & x_1 > 0 \\ B_2(\partial)u(0+, x_2) = \sum_{|\alpha| \leq m_2} b_2^\alpha \partial^\alpha u(0+, x_2) = f_2(x_2), & x_2 > 0 \end{cases}, \tag{7.2}$$

where $a^\alpha, b_l^\alpha \in \mathbb{C}$, and $\partial = (\partial_1, \partial_2)$ is the differentiation. The boundary data $f_l(x_l)$ are tempered distributions on $\mathbb{R}^+$, i.e.,

$$f_l \in S'(R^+) := \{f|_{R^+} : f \in S'(\mathbb{R})\}. \tag{7.3}$$

© Springer Nature Switzerland AG 2019
A. Komech, A. Merzon, *Stationary Diffraction by Wedges*, Lecture Notes in Mathematics 2249, https://doi.org/10.1007/978-3-030-26699-8_7

In particular, the homogeneous Helmholtz equation $[\Delta + \omega^2]u(x) = 0$ for $x \in Q$ in new coordinates reads

$$A(\partial)u(x) = [\Delta - 2\cos\Phi\,\partial_1\partial_2 + \omega^2\sin^2\Phi]u(x) = 0, \qquad x \in K. \qquad (7.4)$$

Denote

$$
\begin{cases}
\tilde{A}(z) := A(-iz) = \sum_{|\alpha|\leq 2} a^\alpha(-iz)^\alpha, & \tilde{A}^0(z) = \sum_{|\alpha|=2} a^\alpha(-iz)^\alpha, \\[2mm]
\tilde{B}_l(z) := B_l(-iz) = \sum_{|\alpha|\leq m_l} b_l^\alpha(-iz)^\alpha, & \tilde{B}_l^0(z) = \sum_{|\alpha|=m_l} b_l^\alpha(-iz)^\alpha
\end{cases}
. \qquad (7.5)
$$

Our basic assumptions are as follows.

I. Strong ellipticity. The operator A is *strongly elliptic*, i.e.

$$|\tilde{A}(z)| \geq \varkappa(|z|^2 + 1), \qquad z \in \mathbb{R}^2, \qquad (7.6)$$

where $\varkappa > 0$. In particular, the polynomial $\tilde{A}(z)$ is *irreducible*, and

$$a^{20} \neq 0, \qquad a^{02} \neq 0. \qquad (7.7)$$

For example, (7.6) holds for the Helmholtz operator (7.4) if $\omega \in \mathbb{C} \setminus \mathbb{R}$.

II. Shapiro-Lopatinski condition. Let us denote the curves

$$C_1^- := \{z \in \mathbb{R} \times \mathbb{C}^- : \tilde{A}^0(z) = 0\}, \qquad C_2^- := \{z \in \mathbb{C}^- \times \mathbb{R} : \tilde{A}^0(z) = 0\},$$

where $\mathbb{C}^\pm := \{z \in \mathbb{C} : \pm\operatorname{Im} z > 0\}$. Then the Shapiro-Lopatinski condition reads

$$\tilde{B}_l^0(z_1, z_2) \neq 0 \quad\text{for}\quad z \in C_l^-, \quad l = 1, 2. \qquad (7.8)$$

For example, (7.8) holds for the Dirichlet, Neumann and Robin b.c.

Let us formulate some important consequences of these conditions. Let us consider the roots λ_l of the characteristic equations

$$\tilde{A}(\lambda_1(z_2), z_2) = 0, \qquad \tilde{A}(z_1, \lambda_2(z_1)) = 0. \qquad (7.9)$$

which are two-valued branching analytic functions on $\mathbb{C}$.

Below we shall need the factorization

$$\tilde{A}(z) = -a^{20}(z_1 - \lambda_1^+(z_2))(z_1 - \lambda_1^-(z_2)) = -a^{02}(z_2 - \lambda_2^+(z_1))(z_2 - \lambda_2^-(z_1)), \qquad z \in \mathbb{R}^2, \qquad (7.10)$$

where $\lambda_l^\pm$ are some branches of the roots. For brevity, let us write the polynomial $\tilde{A}$ as

$$\tilde{A}(z) = az_1^2 + 2bz_1z_2 + cz_2^2 + \mathcal{O}(|z|), \qquad z \in \mathbb{C}^2, \quad |z| \to \infty. \tag{7.11}$$

Then, for example,

$$\lambda_1^\pm(z_2) = -bz_2+\mathcal{O}(1)\pm z_2\sqrt{D + \mathcal{O}(1/|z_2|)} = a_1^\pm z_2+\mathcal{O}(\sqrt{|z_2|}), \qquad z_2 \in \mathbb{C}, \quad |z_2| \to \infty. \tag{7.12}$$

Here $D := b^2 - ac$ and $a_1^\pm := -b \pm \sqrt{D}$, where an arbitrary square root is chosen. Note that $a_1^\pm \neq 0$, for otherwise (7.10) would contradict (7.6) with $z_2 \in \mathbb{R}$ as $|z_2| \to \infty$. Therefore, (7.12) implies asymptotics

$$\lambda_1^\pm(z_2) \sim a_1^\pm z_2, \quad |z_2| \to \infty. \tag{7.13}$$

Let us show that

$$\operatorname{Im} a_1^\pm \neq 0. \tag{7.14}$$

Indeed, the strong ellipticity (7.6) implies that

$$\tilde{A}^0(z) \neq 0, \qquad z \in \mathbb{R}^2 \setminus 0. \tag{7.15}$$

Then the corresponding roots of the equation $\tilde{A}^0(v_1^\pm(z_2), z_2) = 0$ are

$$v_1^\pm(z_2) = a_1^\pm z_2. \tag{7.16}$$

Hence,

$$\tilde{A}^0(v_1^\pm(z_2), z_2) = \tilde{A}^0(a_1^\pm, 1)z_2 = 0, \qquad z_2 \in \mathbb{R}.$$

Now (7.14) follows from (7.15).

Lemma 7.1 *Let condition (7.6) hold. Then we can choose $\pm\operatorname{Im} a_1^\pm > 0$, and there exist branches*

$$\lambda_1^\pm(z_2) \in \mathbb{C}^\pm \quad \text{for} \quad z_2 \in \mathbb{R}, \qquad \text{and} \qquad \lambda_2^\pm(z_1) \in \mathbb{C}^\pm \quad \text{for} \quad z_1 \in \mathbb{R} \tag{7.17}$$

with asymptotics

$$\lambda_1^\pm(z_2) \sim \begin{cases} a_1^\pm z_2, \ z_2 \to +\infty \\ a_1^\mp z_2, \ z_2 \to -\infty \end{cases}, \tag{7.18}$$

and similar asymptotics hold for $\lambda_2^\pm(z_1)$. □

Proof Let us prove that the signs of $\operatorname{Im} a_1^{\pm}$ are different. First, (7.12) implies that for $z_2 \gg 1$ there exists a branch of $\lambda_1(z_2)$ with the asymptotics $\sim a_1^+ z_2$ for $z_2 \to +\infty$, and for $z_2 \ll -1$ there exists a branch with the asymptotics $\sim a_1^+ z_2$ for $z_2 \to -\infty$. Hence, these branches lie in different half-planes $\mathbb{C}^{\pm}$ by (7.14). Therefore, any extensions of these branches to $z_2 \in \mathbb{R}$ lie in different half-planes $\mathbb{C}^{\pm}$ by (7.6).

Let us denote the first branch by $\lambda_1^+(z_2)$, and the second, by $\lambda_1^-(z_2)$. Then the following asymptotics hold:

$$\lambda_1^+(z_2) \sim a_1^+ z_2, \qquad z_2 \to +\infty, \qquad \text{and} \qquad \lambda_1^-(z_2) \sim a_1^+ z_2, \qquad z_2 \to -\infty. \tag{7.19}$$

Now (7.12) implies the asymptotics

$$\lambda_1^+(z_2) \sim a_1^- z_2, \qquad z_2 \to -\infty, \qquad \text{and} \qquad \lambda_1^-(z_2) \sim a_1^- z_2, \qquad z_2 \to +\infty, \tag{7.20}$$

since the asymptotics $\sim a_1^+ z_2$ here cannot hold. Indeed, otherwise, the roots $\lambda_1^+(z_2)$ with $z_2 \ll -1$ and $z_2 \gg 1$ would not lie in the same complex half-plane by (7.14). The same arguments apply to the roots $\lambda_1^-(z_2)$. Now, the asymptotics (7.19) and (7.20) imply that the signs of $\operatorname{Im} a_1^{\pm}$ are different.

Hence, we can assume that $\pm \operatorname{Im} a_1^{\pm} > 0$. Then the asymptotics (7.19)–(7.20) imply (7.18), which, in turn, gives (7.17) by (7.6). $\qquad \square$

Remark 7.2 The key points in the proof are the asymptotics (7.19), (7.20), which follow from (7.13) by (7.14). These asymptotics do not follow from (7.13) with real $a_1^{\pm}$.

Lemma 7.3 *Let conditions (7.6) and (7.8) hold. Then*

$$|\tilde{B}_1(z_1, \lambda_2^-(z_1))| \geq \varkappa_1 |z_1|^{m_1}, \qquad |z_1| > R, \quad z_1 \in \mathbb{R} \tag{7.21}$$

for sufficiently large $R > 0$, where $\varkappa_1 > 0$, and a similar estimate holds for $\tilde{B}_2(\lambda_1^-(z_2), z_2)$. $\qquad \square$

Proof The condition (7.8) can be rewritten as

$$\tilde{B}_1^0(v_1(z_2), z_2) \neq 0 \text{ for } z_2 \in \mathbb{R} \backslash 0, \qquad \text{and} \qquad \tilde{B}_2^0(z_1, v_2(z_1)) \neq 0 \text{ for } z_1 \in \mathbb{R} \backslash 0.$$

Now (7.21) follows from (7.16) and from the asymptotics (7.18) for $\lambda_2^-(z_1)$. $\qquad \square$

We will solve the general boundary problem (7.1)–(7.2) in quadratures. Our method is applicable to arbitrary tempered distributions

$$u \in S'(K) := \{v|_K : v \in S'(R^2)\}. \tag{7.22}$$

We will develop the method simultaneously for the case of "distributional solu-tions" (7.22) and for "regular solutions"

$$u \in H^s(K) := \{v|_K : v \in H^s(\mathbb{R}^2)\} \quad \text{with} \quad s > \max(3/2, m_1 + 1/2, m_2 + 1/2).$$
$$(7.23)$$

Our plan is as follows. First, we will express the problem (7.1)–(7.2) in terms of the Cauchy data of the solution as the system of algebraic equations on the Riemann surface of complex characteristics of the strongly elliptic operator $A(\partial)$. Then we will reduce this system to the Riemann–Hilbert problem on the Riemann surface, which will be solved in quadratures.

Chapter 8
Extension to the Plane

We extend Eq. (7.1) to the whole plane which is necessary for application of the Fourier transform.

8.1 Regular Solutions

For regular solutions (7.23) we can extend the solution by zero outside K:

$$u_0(x) = \begin{cases} u(x), & x \in K \\ 0, & x \in \mathbb{R}^2 \setminus K \end{cases}. \tag{8.1}$$

Then (7.1) implies the equation on the whole plane

$$Au_0(x) = \gamma(x), \qquad x \in \mathbb{R}^2. \tag{8.2}$$

Here $\gamma(x)$ is, due to (7.23), a sum of simple and double layers,

$$\gamma(x) = \sum_{0 \le k \le 1} \gamma_1^k(x_1)\delta^{(k)}(x_2) + \sum_{0 \le k \le 1} \gamma_2^k(x_2)\delta^{(k)}(x_1), \qquad x \in \mathbb{R}^2, \tag{8.3}$$

where

$$\gamma_l^k(x_l) \in S'(\overline{\mathbb{R}^+}) := \{g(x) \in S'(\mathbb{R}) : \operatorname{supp} g \subset \overline{\mathbb{R}^+}\}.$$

These distributions are not uniquely defined: for example, (8.3) also holds with $\gamma_1^0(x_1) - \delta(x_1)$ and $\gamma_2^0(x_2) + \delta(x_2)$ instead of $\gamma_1^0(x_1)$ and $\gamma_2^0(x_2)$.

© Springer Nature Switzerland AG 2019
A. Komech, A. Merzon, *Stationary Diffraction by Wedges*, Lecture Notes in Mathematics 2249, https://doi.org/10.1007/978-3-030-26699-8_8

Applying the Fourier transform in two variables to (8.2), we obtain

$$\tilde{A}(z)\tilde{u}_0(z) = \tilde{\gamma}(z), \qquad z \in \mathbb{R}^2, \tag{8.4}$$

where the Fourier transform for test functions $\varphi(x) \in C_0^\infty(R^2)$ is defined by

$$F\varphi(x) = \tilde{\varphi}(\xi) := \int_{\mathbb{R}^2} e^{i\xi x} \varphi(x) dx.$$

Hence, using the strong ellipticity (7.6), we can divide by $\tilde{A}(z)$ and obtain the formula

$$u_0(x) = F^{-1} \frac{\tilde{\gamma}(z)}{\tilde{A}(z)}, \qquad x \in \mathbb{R}^2; \qquad u(x) = u_0(x), \quad x \in K. \tag{8.5}$$

Thus, to find the solution, it remains to calculate the distribution $\gamma(x)$ in terms of the boundary data f_l. This calculation is the main goal of our approach.

First, we will express the distribution $\gamma(x)$ in the Cauchy data of the solution u. Namely, the Sobolev theorem on traces implies that for regular solutions (7.23) there exist the Cauchy data

$$\left\{ \begin{array}{ll} u_1^\beta(x_1) := \partial_2^\beta u_0(x_1, 0+), & x_1 > 0 \\[2mm] u_2^\beta(x_2) := \partial_2^\beta u_0(0+, x_2), & x_2 > 0 \end{array} \right|, \quad \beta = 0, 1, \tag{8.6}$$

where the limits hold in $H^{s-\beta-1/2}(\mathbb{R}^+)$, and $u_l^\beta \in H^{s-\beta-1/2}(\mathbb{R}^+)$. Let us introduce notations

$$A(\partial) = \sum_j A_{1j}(\partial_1)\partial_2^j = \sum_j A_{2j}(\partial_2)\partial_1^j, \tag{8.7}$$

where

$$\left\{ \begin{array}{lll} A_{10}(\partial_1) = a^{20}\partial_1^2 + a^{10}\partial_1 + a^{00}, & A_{11}(\partial_1) = a^{11}\partial_1 + a^{01}, & A_{12}(\partial_1) = a^{02} \\[2mm] A_{20}(\partial_2) = a^{02}\partial_2^2 + a^{01}\partial_2 + a^{00}, & A_{21}(\partial_2) = a^{11}\partial_2 + a^{10}, & A_{22}(\partial_1) = a^{20} \end{array} \right|. \tag{8.8}$$

Then applying the differential operator $A(\partial)$ to the discontinuous function u_0 we obtain (8.3), where

$$\left\{ \begin{array}{ll} \gamma_1^k(x_1) = \sum_{j \geq k+1} A_{1j}(\partial_1)u_1^{j-k-1}(x_1), & x_1 > 0 \\[2mm] \gamma_2^k(x_2) = \sum_{j \geq k+1} A_{2j}(\partial_2)u_2^{j-k-1}(x_2), & x_2 > 0 \end{array} \right|, \quad k = 0, 1. \tag{8.9}$$

However, these formulas do not specify the distributions $\gamma_l^k \in S'(\overline{\mathbb{R}^+})$ uniquely.

Let us denote by $v_l^\beta \in S'(\overline{\mathbb{R}^+})$ some extensions of the Cauchy data by zero for negative arguments, i.e.,

$$v_l^\beta(x_l) = \begin{cases} u_l^\beta(x_l), & x_l > 0 \\ 0, & x_l < 0 \end{cases}, \qquad l = 1, 2, \quad \beta = 0, 1. \tag{8.10}$$

Lemma 8.1 *There exist extensions* $v_l^\beta \in S'(\overline{\mathbb{R}^+})$ *of the Cauchy data such that*

$$\begin{cases} \gamma_1^k(x_1) = \sum_{j \geq k+1} A_{1j}(\partial_1) v_1^{j-k-1}(x_1), & x_1 \in \mathbb{R} \\ \gamma_2^k(x_2) = \sum_{j \geq k+1} A_{2j}(\partial_2) v_2^{j-k-1}(x_2), & x_2 \in \mathbb{R} \end{cases}, \qquad k = 0, 1. \tag{8.11}$$

Proof (8.9) Implies that, for arbitrary extensions $v_l^\beta \in S'(\overline{\mathbb{R}^+})$, the identities (8.11) hold up to some distributions supported by the point $x_l = 0$. For example, the first two identities read

$$\gamma_1^0(x_1) = (a^{11}\partial_1 + a^{01})v_1^0(x_1) + a^{02}v_1^1(x_1) + \sum_j C_j \delta^{(j)}(x_1), \quad \gamma_1^1(x_1) = a^{02}v_1^0(x_1) + \sum_j D_j \delta^{(j)}(x_1), \qquad x_1 \in \mathbb{R}.$$

Finally, these point distributions can be included into v_1^1 and v_1^0 by (7.7). □

8.2 Distributional Solutions

Now let us justify all relations (8.3), (8.6) and (8.11) for distributional solutions (7.22). In this case the extension $u_0 \in S'(R^2)$ of u by zero outside K exists, though it is not unique, and (8.3) becomes

$$\gamma(x) := A(\partial)u_0(x) = \sum_{0 \leq k \leq N} \gamma_1^k(x_1)\delta^{(k)}(x_2) + \sum_{0 \leq k \leq N} \gamma_2^k(x_2)\delta^{(k)}(x_1), \qquad x \in \mathbb{R}^2, \tag{8.12}$$

where $\gamma_l^k \in S'(\overline{\mathbb{R}^+})$, and $N < \infty$ is bounded by the order of singularity of the distribution u. Let us show that the number N can be reduced to $N = 2$ by using the Cauchy-Kowalevski method.

Lemma 8.2 *Let* $u \in S'(K)$ *be a distributional solution to Eq. (7.1). Then there exists an extension by zero* $u_0 \in S'(\mathbb{R}^2)$ *such that (8.3) holds with* $\gamma_l^k \in S'(\overline{\mathbb{R}^+})$.

Proof Applying the Fourier transform to (8.12), we obtain

$$\tilde{A}(z)\tilde{u}_0(z) = \tilde{\gamma}(z) = \sum_{0 \le k \le N} \tilde{\gamma}_1^k(z_1)(-iz_2)^k + \sum_{0 \le k \le N} \tilde{\gamma}_2^k(z_2)(-iz_1)^k, \qquad z \in \mathbb{R}^2.$$

(8.13)

Divide the first sum in (8.13) by the symbol $\tilde{A}(z_1, z_2)$ in the modulus of polynomials of z_2 with coefficients from $S'(\mathbb{R})$. Similarly, we divide the second sum in (8.13) by the symbol $\tilde{A}(z_1, z_2)$ in the modulus of polynomials of z_1 with coefficients from $S'(\mathbb{R})$. Then (7.7) implies that

$$\tilde{A}(z)\tilde{u}_0(z) = \tilde{A}(z)\tilde{Q}(z) + \sum_{0 \le k \le 1} \tilde{r}_1^k(z_1)(-iz_2)^k + \sum_{0 \le k \le 1} \tilde{r}_2^k(z_2)(-iz_1)^k, \qquad z \in \mathbb{R}^2,$$

(8.14)

where $Q(x)$ is a distribution of the form (8.12), and $r_1^k, r_2^k \in S'(\overline{\mathbb{R}})$. It remains to define the modified extension as $u_0 - Q$. □

Lemma 8.3 *Let $u \in S'(K)$ be a distributional solution to (7.1). Then the Cauchy data (8.6) exist, where the limits hold in $S'(\mathbb{R}^+)$.* □

Proof (i) The representation (8.3), which was proved in previous lemma, implies that

$$u_0(x) = F^{-1}\frac{\tilde{\gamma}(z)}{\tilde{A}(z)} = F^{-1}\frac{\sum_{0 \le k \le 1}\tilde{\gamma}_1^k(z_1)(-iz_2)^k + \sum_{0 \le k \le 1}\tilde{\gamma}_2^k(z_2)(-iz_1)^k}{\tilde{A}(z)}, \qquad x \in K.$$

(8.15)

It suffices to prove the existence of the Cauchy data for each term

$$w(x) = F^{-1}\frac{\tilde{\gamma}_2^k(z_2)(-iz_1)^k}{\tilde{A}(z)}, \qquad x \in K$$

(8.16)

with $k = 0, 1$. This distribution is a convolution of the layer $\gamma_2^k(x_2)\delta^{(k)}(x_1)$ with the fundamental solution $E(x) = F^{-1}[1/\tilde{A}(z)]$. The function $1/\tilde{A}(z)$ is analytic for $|\text{Im } z| < \varepsilon$ with small $\varepsilon > 0$ by strong ellipticity (7.6). Hence, the fundamental solution $E(x)$ is smooth for $x \ne 0$, and it decays exponentially at infinity by the Paley-Wiener theorem [115, 122]. Hence, $w(x)$ is a smooth function in a neighborhood of the semi-axis $x_2 = 0, x_1 > 0$.

It remains to prove the existence of the Cauchy data on the axis $x_1 = 0$. This follows from the "theory of transmission" of Vishik and Eskin [14, 141]. We recall their arguments for the convenience of the reader. In the notation of (7.10)

$$\frac{1}{z_1 - \lambda_1^-(z_2)} = \frac{1}{z_1 - i + i - \lambda_1^-(z_2)} = \frac{1}{z_1 - i}\frac{1}{1 + \frac{i - \lambda_1^-(z_2)}{z_1 - i}}$$

(8.17)

$$= \frac{1}{z_1 - i}\left[1 + \mathcal{O}\left(\frac{i - \lambda_1^-(z_2)}{z_1 - i}\right)\right], \qquad z \in \mathbb{R}^2.$$

This asymptotics holds, since the quotation $\frac{i - \lambda_1^-(z_2)}{z_1 - i}$ is bounded away from -1 for $z \in \mathbb{R}^2$. Indeed,

$$|1 + \frac{i - \lambda_1^-(z_2)}{z_1 - i}| = |\frac{z_1 - \lambda_1^-(z_2)}{z_1 - i}| \geq \delta, \qquad z \in \mathbb{R}^2, \tag{8.18}$$

where $\delta > 0$. Equivalently, $|z_1 - \lambda_1^-(z_2)| \geq \delta|z_1 - i|$, which holds by the asymptotics (7.18) since $|z_1 - \lambda_1^-(z_2)|$ is greater than the distance from the point $z_1 \in \mathbb{R}$ to the curve $\{\lambda_1^-(z_2) : z_2 \in \mathbb{R}\}$.

Using (8.17) we obtain that

$$\tilde{w}(z) = -\frac{\tilde{\gamma}_2^k(z_2)(-iz_1)^k}{a^{20}(z_1 - \lambda_1^+(z_2))(z_1 - i)}$$

$$-\frac{\tilde{\gamma}_2^k(z_2)(-iz_1)^k}{a^{20}(z_1 - \lambda_1^+(z_2))(z_1 - i)}\mathcal{O}\left(\frac{i - \lambda_1^-(z_2)}{z_1 - i}\right), \qquad z \in \mathbb{R}^2. \tag{8.19}$$

The first term on the right-hand side is analytic in $\operatorname{Im} z_1 < 0$, and hence, its inverse Fourier transform vanishes for $x_1 > 0$ by the Paley-Wiener theorem [115, 122]. On the other hand, the second term is summable as a function of $z_1 \in \mathbb{R}$ with values in $S'(\mathbb{R})$, because

$$\int \left|\frac{(-iz_1)^k}{(z_1 - \lambda_1^+(z_2))(z_1 - i)}\mathcal{O}\left(\frac{i - \lambda_1^-(z_2)}{z_1 - i}\right)\right| dz_1$$

$$\leq C|(i - \lambda_1^-(z_2))| \int \frac{|z_1|^k}{|z_1 - i|^3} dz_1 \leq C_1(1 + |z_2|), \qquad z_2 \in \mathbb{R} \tag{8.20}$$

for $k = 0, 1$. Hence, the inverse Fourier transform of the second term on the right-hand side of (8.19) is a continuous function of $x_1 \in \mathbb{R}$ with values in $S'(\mathbb{R})$ by (7.18). $\qquad \square$

Lemma 8.4 *Let $u \in S'(K)$ be a distributional solution to (7.1) represented by (8.5), where γ is given by (8.3), (8.11) with some distributions $v_l^\beta \in S'(\overline{\mathbb{R}^+})$. Then the Cauchy data of u satisfy*

$$u_l^\beta = v_l^\beta|_{\mathbb{R}^+}, \qquad l = 1, 2, \quad \beta = 0, 1. \tag{8.21}$$

Proof Equation (8.5) implies that $u_0 = u_1 + u_2$, where

$$u_1(x) := F^{-1} \frac{\sum_{0 \le k \le 1} \tilde{\gamma}_1^k(z_1)(-iz_2)^k}{\tilde{A}(z)}, \qquad u_2(x) := F^{-1} \frac{\sum_{0 \le k \le 1} \tilde{\gamma}_2^k(z_2)(-iz_1)^k}{\tilde{A}(z)}, \qquad x \in \mathbb{R}^2.$$
(8.22)

It suffices to check the identities (8.21) with $l = 1$. They follow from the jumps

$$u_0(x_1, 0+) - u_0(x_1, 0-) = v_1^0(x_1), \qquad \partial_2 u_0(x_1, 0+) - \partial_2 u_0(x_1, 0-) = v_1^1(x_1), \qquad x_1 > 0,$$
(8.23)

since $u_0(x) = 0$ outside $\overline{K}$. These jumps vanish for u_2, since this distribution is a smooth function on $\mathbb{R}^+ \times 0$. Hence, it suffices to check these jumps for u_1. Let us check, moreover, that

$$u_1(x_1, 0+) - u_1(x_1, 0-) = v_1^0(x_1), \qquad \partial_2 u_1(x_1, 0+) - \partial_2 u_1(x_1, 0-) = v_1^1(x_1), \qquad x_1 \in \mathbb{R}.$$
(8.24)

Let us denote by

$$\hat{E}(z_1, x_2) = F_{z_2 \to x_2}^{-1} \frac{1}{\tilde{A}(z_1, z_2)}$$

the fundamental solution of 1D operator $A(-iz_1, \partial_2)$: for $z_1 \in \mathbb{R}$

$$A(-iz_1, \partial_2)\hat{E}(z_1, x_2) = \delta(x_2), \qquad x_2 \in \mathbb{R}. \tag{8.25}$$

Then $\hat{u}_1(z_1, x_2) := F_{x_1 \to z_1} u_1$ reads as

$$\hat{u}_1(z_1, x_2) = \sum_{0 \le k \le 1} \tilde{\gamma}_1^k(z_1) \partial_2^k \hat{E}(z_1, x_2), \qquad (z_1, x_2) \in \mathbb{R}^2.$$

Hence,

$$\hat{u}_1(z_1, 0+) - \hat{u}_1(z_1, 0-) = \tilde{\gamma}_1^1(z_1)[\partial_2 \hat{E}(z_1, 0+) - \partial_2 \hat{E}(z_1, 0-)], \qquad (z_1, x_2) \in \mathbb{R}^2,$$

since $\hat{E}(z_1, x_2)$ is a continuous function. Finally,

$$\partial_2 \hat{E}(z_1, 0+) - \partial_2 \hat{E}(z_1, 0-) = 1/a^{02}, \tag{8.26}$$

while $\gamma_1^1(x_1) = a^{02} v_1^0(x_1)$ by (8.11) and (8.7). This proves the first jump in (8.24). The second jump in (8.24) requires more attention:

$$\partial_2 \hat{u}_1(z_1, 0+) - \partial_2 \hat{u}_1(z_1, 0-) = \sum_{0 \le k \le 1} \tilde{\gamma}_1^k(z_1)[\partial_2^{k+1} \hat{E}(z_1, 0+) - \partial_2^{k+1} \hat{E}(z_1, 0-)], \qquad z_1 \in \mathbb{R}.$$
(8.27)

The jump of $\partial_2^2 \hat{E}(z_1, x_2)$ can be calculated from Eq. (8.25): in notations (8.7), (8.72)

$$a^{02} \partial_2^2 \hat{E}(z_1, x_2) + [a^{11}(-iz_1) + a^{01}] \partial_2 \hat{E}(z_1, x_2)$$

$$+ [a^{20}(-iz_1)^2 + a^{10}(-iz_1) + a^{00}] \hat{E}(z_1, x_2) = \delta(x_2), \qquad x_2 \in \mathbb{R}.$$

Hence, (8.26) implies the jump

$$\partial_2^2 \hat{E}(z_1, 0+) - \partial_2^2 \hat{E}(z_1, 0-) = -\frac{1}{(a^{02})^2}[a^{11}(-iz_1) + a^{01}] \qquad (8.28)$$

since $\hat{E}(z_1, x_2)$ is continuous at $x_2 = 0$. Now, substituting (8.28) and (8.26) into (8.27), we obtain

$$\partial_2 \hat{u}_1(z_1, 0+) - \partial_2 \hat{u}_1(z_1, 0-) = \frac{1}{a^{02}}\tilde{\gamma}_1^0(z_1) - \frac{1}{(a^{02})^2}[a^{11}(-iz_1) + a^{01}]\tilde{\gamma}_1^1(z_1), \qquad z_1 \in \mathbb{R}.$$
$$(8.29)$$

On the other hand, (8.11) and (8.8) imply that

$$\gamma_1^0 = A_{11}(\partial_1)v_1^0 + A_{12}v_1^1 = (a^{11}\partial_1 + a^{01})v_1^0 + a^{02}v_1^1, \qquad \gamma_1^1(x_1) = A_{12}v_1^0 = a^{02}v_1^0.$$

Substituting into (8.29), we finally obtain

$$\partial_2 \hat{u}_1(z_1, 0+) - \partial_2 \hat{u}_1(z_1, 0-)$$

$$= \frac{1}{a^{02}}[(a^{11}(-iz_1) + a^{01})v_1^0 + a^{02}v_1^1] - \frac{1}{(a^{02})^2}[a^{11}(-iz_1) + a^{01}]a^{02}v_1^0 = v_1^1(z_1), \qquad z_1 \in \mathbb{R}.$$

Now the second jump in (8.24) is also proven.   □

Remark 8.5 The function $u_1(x)$ is the sum of potentials of simple and double layers corresponding to terms in the sums (8.22) with $k = 0$ and $k = 1$, respectively. The relations (8.24) extend classical formulas for the jumps of the Coulombic potentials and their normal derivatives to the setting of general strongly elliptic operators.   □

Chapter 9
Boundary Conditions via the Cauchy Data

To construct all solutions to the problem (7.1), (7.2) by formula (8.5), we need to calculate all "extended Cauchy data" $v_l^\beta \in S'(\overline{\mathbb{R}^+})$, i.e., four functions of one variable. This is the main goal of our approach. Thus, we need at least four equations. Two of the equations follow from the boundary conditions (7.2). The third equation will be extracted from the identity (8.4). The needed additional equation we will obtain by the Malyshev method of automorphic functions [77].

Here we obtain two equations for the Cauchy data (8.6) from the boundary conditions (7.2). For example, boundary conditions are just expressed via the Cauchy data in the simplest case of first-order boundary operators, e.g., for the Dirichlet or Neumann b.c. For higher order boundary operators this is also possible by the Cauchy-Kowalevski method, since the elliptic operator $A(\partial)$ does not have real characteristics. Let us denote by $f_l^0 \in S'(\mathbb{R}^+)$ some extensions of distributions f_l by zero for negative arguments.

Lemma 9.1

(i) *Let the strong ellipticity condition (7.6) hold. Then boundary conditions (7.2) for distributions $f_l^0 \in S'(\mathbb{R}^+)$ are equivalent to identities*

$$\begin{cases} B_{10}(\partial_1)v_1^0(x_1) + B_{11}(\partial_1)v_1^1(x_1) = f_1^0(x_1) + \sum C_{1j}\delta^{(j)}(x_1), & x_1 \in \mathbb{R} \\ B_{20}(\partial_1)v_2^0(x_2) + B_{21}(\partial_1)v_2^1(x_2) = f_2^0(x_2) + \sum C_{2j}\delta^{(j)}(x_2)), & x_2 \in \mathbb{R} \end{cases},$$
(9.1)

where $B_{l\beta}$ are some differential operators with constant coefficients, $\{C_{lj}\}$ is a finite set of complex numbers, and $v_l^\beta \in S'(\overline{\mathbb{R}^+})$ are some extensions (8.10).

(ii) *Let, in addition, the Shapiro-Lopatinski condition (7.8) hold. Then, for each $l = 1, 2$,*

$$\deg B_{l0} = m_l \qquad or \qquad \deg B_{l1} = m_l - 1. \tag{9.2}$$

© Springer Nature Switzerland AG 2019
A. Komech, A. Merzon, *Stationary Diffraction by Wedges*, Lecture Notes in Mathematics 2249, https://doi.org/10.1007/978-3-030-26699-8_9

Proof

(i) Dividing the polynomials $B_l(\partial)$ by $A(\partial)$, we obtain

$$B_1(\partial) = A(\partial)q_1(\partial) + C_1(\partial), \qquad B_2(\partial) = A(\partial)q_2(\partial) + C_2(\partial). \qquad (9.3)$$

Here q_1 and q_2 are the corresponding quotients, and C_1 and C_2 are the remainders, which read as

$$C_1(\partial) = B_{10}(\partial_1) + B_{11}(\partial_1)\partial_2, \qquad C_2(\partial) = B_{20}(\partial_2) + B_{21}(\partial_2)\partial_1 \qquad (9.4)$$

by (7.6). Hence, boundary conditions (7.2) become (9.1) due to (7.1) and (8.10). Conversely, (9.1) implies (7.2) by (9.3) and (9.4).

(ii) Equations (7.21) and (9.4) imply that

$$|\tilde{C}_1(z_1, \lambda_2^-(z_1))| = |\tilde{B}_1(z_1, \lambda_2^-(z_1))| \geq \varkappa_1 |z_1|^{m_1}, \qquad |z_1| > R, \quad z_1 \in \mathbb{R}. \qquad (9.5)$$

Hence,

$$|\tilde{B}_{10}(z_1) + \tilde{B}_{11}(z_1)(-i\lambda_2^-(z_1))| \geq \varkappa_1 |z_1|^{m_1}, \qquad |z_1| > R, \quad z_1 \in \mathbb{R} \qquad (9.6)$$

by (9.3). This implies (9.2) with $l = 1$. $\square$

Corollary 9.2 *Equations (9.3) and (9.4) imply that*

$$\left\{ \begin{array}{l} B_{10}(\partial_1) + B_{11}(\partial_1)\partial_2 = B_1(\partial) \bmod A(\partial) \\[1em] B_{20}(\partial_2) + B_{21}(\partial_2)\partial_1 = B_2(\partial) \bmod A(\partial) \end{array} \right|, \qquad (9.7)$$

which is well known in the theory of PDEs [3, 8].

Let us recall that the complex Fourier transform $\tilde{v}_l^\beta(z_l)$, $\tilde{f}_l^0(z_l)$ of the tempered distributions $v_l^\beta(x_l)$, $f_l^0(x_l) \in S'(\mathbb{R}^+)$ are analytic in $\mathrm{Im}\, z_l > 0$ by the Paley–Wiener theorem [54, 115, 122] and they satisfy the estimates

$$|\tilde{v}_l^\beta(z_l)| \leq C \frac{(|z_l| + 1)^M}{|\mathrm{Im}\, z_l|^N}, \qquad |\tilde{f}_l^0(z_l)| \leq C \frac{(|z_l| + 1)^M}{|\mathrm{Im}\, z_l|^N}, \qquad \mathrm{Im}\, z_l > 0, \qquad (9.8)$$

where M characterizes the order of singularity of the corresponding distributions $v_l^\beta(x_l)$ and $f_l^0(x_l)$, while N corresponds to their growth at infinity. Conversely, this analyticity and the estimates imply that $v_l^\beta(x_l)$, $f_l^0(x_l) \in S'(\mathbb{R}^+)$, see [54, Theorem 5.2].

The boundary conditions (9.1) after the complex Fourier transform give two algebraic equations for four unknown functions $\tilde{v}_l^\beta$:

$$
\begin{cases}
\tilde{B}_{10}(z_1)\tilde{v}_1^0(z_1) + \tilde{B}_{11}(z_1)\tilde{v}_1^1(z_1) = g_1(z_1) := \tilde{f}_1^0(z_1) + \sum C_{1j}(-iz_1)^j, & \operatorname{Im} z_1 > 0, \\[2ex]
\tilde{B}_{20}(z_2)\tilde{u}_2^0(z_2) + \tilde{B}_{21}(z_2)\tilde{v}_2^1(z_2) = g_2(z_2) := \tilde{f}_2^0(z_2) + \sum C_{2j}(-iz_2)^j, & \operatorname{Im} z_2 > 0.
\end{cases}
$$

$$(9.9)$$

Chapter 10
Connection Equation on the Riemann Surface

Here we obtain the third equation for the Cauchy data using the identity (8.4). One of the central ideas of our method is that this identity also holds in the complex region, in the *tube domain* $\mathbb{C}K^* \subset \mathbb{C}^2$ defined by

$$\mathbb{C}K^* := \{z \in \mathbb{C}^2 : \operatorname{Im} z_1 > 0, \operatorname{Im} z_2 > 0\}. \qquad (10.1)$$

Namely, the Paley-Wiener theorem [115, 122] (see also [54, Theorem 5.2]) implies that distributions $\tilde{u}_0(z)$ and $\tilde{\gamma}(z)$ on the real plane $\mathbb{R}^2$ are traces of analytic functions in $\mathbb{C}K^*$. Hence,

$$\tilde{A}(z)\tilde{u}_0(z) = \tilde{\gamma}(z), \qquad z \in \mathbb{C}K^*. \qquad (10.2)$$

Alternatively, this identity follows from (8.2) by complex Fourier-Laplace transform, i.e., by scalar product of both sides with the "test function" e^{izx}.

Definition 10.1 The Riemann surface of complex characteristics of the elliptic operator $A(\partial)$ is defined by

$$V := \{z \in \mathbb{C}^2 : \tilde{A}(z) = 0\}. \qquad (10.3)$$

Let us denote

$$V^* := V \cap \mathbb{C}K^*, \qquad (10.4)$$

see Fig. 10.1. This intersection is nonempty. Indeed, (7.17) implies that this intersection contains, in particular, the points $(z_1 + i\varepsilon, \lambda_2^+(z_1 + i\varepsilon))$ with $z_1 \in \mathbb{R}$ and small $\varepsilon > 0$.

© Springer Nature Switzerland AG 2019
A. Komech, A. Merzon, *Stationary Diffraction by Wedges*, Lecture Notes in Mathematics 2249, https://doi.org/10.1007/978-3-030-26699-8_10

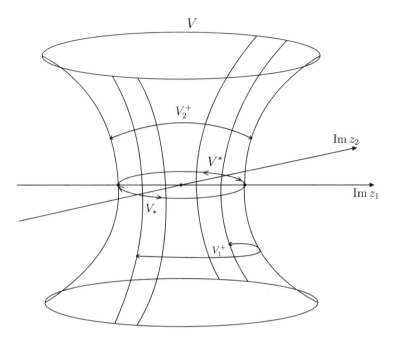

Fig. 10.1 Riemann surface of complex characteristics

Connection Equation The identity (10.2) implies the key "Connection Equation" on the Riemann surface:

$$\tilde{\gamma}(z) = 0, \qquad z \in V^*. \tag{10.5}$$

Remark 10.2 The identity (10.5) extends the well known relations between the Cauchy data on the real characteristics of hyperbolic PDEs, since $\gamma(x)$ is expressed via the Cauchy data by (8.3) and (8.11).   □

The following theorem follows from general Division Theorem [53, 87], since the polynomial $\tilde{A}(z)$ is irreducible.

Theorem 10.3 *Let $A(\partial)$ by a strongly elliptic operator, and $\tilde{\gamma}(z)$ be a Fourier–Laplace transform of a tempered distribution $\gamma \in S'(\overline{K})$, satisfying the Connection Equation (10.5). Then the quotient*

$$\tilde{u}_0(z) := \frac{\tilde{\gamma}(z)}{\tilde{A}(z)}, \qquad z \in \mathbb{C}K^* \setminus V \tag{10.6}$$

is the Fourier–Laplace transform of a unique tempered distribution $u_0 \in S'(\overline{K})$. □

Proof This quotient is an analytic function in $\mathbb{C}K^*$ since the polynomial $\tilde{A}(z)$ is irreducible by (7.6). Hence, this theorem follows from general Division Theorem [53, 87].   □

Remark 10.4 The key role of complex zeros of symbol (10.3) was suggested by the theory of boundary value problems in the half-plane $x_1 > 0$, where exponentially decaying solutions $e^{-iz_1(z_2)x_1}$ correspond to "stable roots" $z_1(z_2)$ of the characteristic equation $\tilde{A}(z_1, z_2) = 0$ with $\operatorname{Im} z_1(z_2) < 0$. Thus, in the half-plane $x_1 > 0$ the "unstable roots" with $\operatorname{Im} z_1 > 0$ are forbidden, and in the half-plane $x_2 > 0$ the unstable roots with $\operatorname{Im} z_2 > 0$ are forbidden. Our condition (10.5) excludes the roots that are forbidden both in the direction $x_1 > 0$ and in $x_2 > 0$, and also in all intermediate directions, corresponding to complex roots $(z_1, z_2) \in \mathbb{C}^2$ of the characteristic equation $\tilde{A}(z_1, z_2) = 0$ with $\operatorname{Im} z_1 > 0$, $\operatorname{Im} z_2 > 0$. □

Remark 10.5 For the problem (7.1), (7.2) in the angle Q of magnitude $\Phi > \pi$, the corresponding tube domain $\mathbb{C}Q^*$ is empty. In this case we have found the suitable generalization of the relation (10.5) in [58]. Namely, a similar relation holds on $V^* := V \cap \mathbb{C}K^*$ where $K := \mathbb{R}^2 \setminus \overline{Q}$ for the *analytic continuations* of the functions $\tilde{v}_l^\beta$ along the Riemann surface V. □

Substituting (8.11) into (8.3) we rewrite (10.5) as

$$\tilde{\gamma}(z) = \sum_{\beta=0,1} A_1^\beta(z) \tilde{v}_1^\beta(z_1) + \sum_{\beta=0,1} A_2^\beta(z) \tilde{v}_2^\beta(z_2) = 0, \quad z \in V^*, \tag{10.7}$$

where polynomials $A_l^\beta(z)$ can be easily expressed in terms of the coefficients of the operator $A(\partial)$. Namely, (8.11) implies that

$$\tilde{\gamma}(z) = \sum_{0 \le k \le 1} \tilde{\gamma}_1^k(z_1)(-iz_2)^k + \sum_{0 \le k \le 1} \tilde{\gamma}_2^k(z_2)(-iz_1)^k$$

$$= \sum_{0 \le k \le 1} \sum_{j \ge k+1} \tilde{A}_{1j}(z_1) \tilde{v}_1^{j-k-1}(z_1)(-iz_2)^k$$

$$+ \sum_{0 \le k \le 1} \sum_{j \ge k+1} \tilde{A}_{2j}(z_2) \tilde{v}_2^{j-k-1}(z_2)(-iz_1)^k, \qquad z \in \mathbb{R}^2. \tag{10.8}$$

Hence,

$$\left\{ \begin{array}{l} A_1^\beta(z) = \displaystyle\sum_{j \ge \beta+1} \tilde{A}_{1j}(z_1)(-iz_2)^{j-1-\beta} \\[4mm] A_2^\beta(z) = \displaystyle\sum_{j \ge \beta+1} \tilde{A}_{2j}(z_2)(-iz_1)^{j-1-\beta} \end{array} \right. . \tag{10.9}$$

Therefore, (8.8) gives

$$\left\{ \begin{array}{l} A_1^1 = a^{02}, \; A_1^0(z) = a^{02}(-iz_2) + a^{11}(-iz_1) + a^{01} \\[2mm] A_2^1 = a^{20}, \; A_2^0(z) = a^{20}(-iz_1) + a^{11}(-iz_2) + a^{10} \end{array} \right. . \tag{10.10}$$

Chapter 11
On Equivalence of the Reduction

We have deduced the identity (10.7) for the Cauchy data of arbitrary distributional solution to Eq. (7.1). Now let us show that the identity (10.7) provides a solution to (7.1) with these Cauchy data.

11.1 Distributional Solutions

Lemma 11.1 *Let the operator A be strongly elliptic, and let identity (10.7) hold for some distributions $v_l^\beta \in S'(\overline{\mathbb{R}^+})$. Then*

(i) *Formulas (8.5) and (8.3), (8.11) define a unique tempered distribution $u_0 \in S'(\mathbb{R}^2)$ with supp $u_0 \subset \overline{K}$.*

(ii) *The distribution $u := u_0|_K \in S'(K)$ is a solution to (7.1), and $u_l^\beta = v_l^\beta|_{\mathbb{R}^+}$ are the Cauchy data of u.* □

Proof Assertion (i) follows from Theorem 10.3, and (ii) follows from Lemma 8.4. □

11.2 Regular Solutions

Let us recall that $u_l^\beta \in H^{s-\beta-1/2}(\mathbb{R}^+)$. In the case $u_l^\beta \in \mathring{H}^{s-\beta-1/2}(\mathbb{R}^+)$ we can choose the corresponding extensions by zero $v_l^\beta \in H^{s-\beta-1/2}(\mathbb{R})$. Lemma A.1 implies that

$$\dim H^s(\mathbb{R}^+)/\mathring{H}^s(\mathbb{R}^+) = [s+1/2] \quad \text{for} \quad s > -1/2 \quad \text{if} \quad s \notin \mathbb{N} - 1/2 \quad (11.1)$$

© Springer Nature Switzerland AG 2019
A. Komech, A. Merzon, *Stationary Diffraction by Wedges*, Lecture Notes
in Mathematics 2249, https://doi.org/10.1007/978-3-030-26699-8_11

(for $|s| < 1/2$ this is the well known Stein's result [134]). Hence, the following lemma means that identity (10.7) is equivalent to Eq. (7.1) for the Cauchy data from some linear subspace of finite codimension.

Lemma 11.2 *Let the operator A be strongly elliptic, let identity (10.7) hold for some distributions $v_l^\beta \in \mathring{H}^{s-\beta-1/2}(\mathbb{R}^+)$ with $s > 3/2$, and let the distribution u_0 be defined by (8.5) and (8.3), (8.11). Then*

(i) $u := u_0|_K \in H^s(K)$, and

$$\|u\|^2_{H^s(K)} \le C \sum_{l,\beta} \|v_l^\beta\|_{\mathring{H}^{s-\beta-1/2}(\mathbb{R}^+)}. \tag{11.2}$$

(ii) *The function u is a solution to (7.1), and $u_l^\beta(x_l) = v_l^\beta(x_l)|_{\mathbb{R}^+}$ are the Cauchy data of u.*   □

Proof

(i) For example, let us estimate the Sobolev norm of the term (8.16) for the case $s \approx 3/2$. The required estimate follows by the "theory of transmission" of Vishik and Eskin [14, 141]. In (8.17), (8.20) we used a simplified version of this theory, but now we need its original version. Namely, the asymptotics (8.17) changes to

$$\frac{1}{z_1 - \lambda_1^-(z_2)} = \frac{1}{z_1 - i(|z_2| + 1)} \frac{1}{1 + \dfrac{i(|z_2| + 1) - \lambda_1^-(z_2)}{z_1 - i(|z_2| + 1)}} \tag{11.3}$$

$$= \frac{1}{z_1 - i(|z_2| + 1)} \left[1 - \frac{i(|z_2| + 1) - \lambda_1^-(z_2)}{z_1 - i(|z_2| + 1)} + \mathcal{O}\left(\left| \frac{i(|z_2| + 1) - \lambda_1^-(z_2)}{z_1 - i(|z_2| + 1)} \right|^2 \right) \right],$$

which follows by the same arguments (8.18). Hence, the Fourier transform of term (8.16) reads

$$\tilde{w}(z) = -\frac{\tilde{\gamma}_2^k(z_2)(-iz_1)^k}{a^{20}(z_1 - \lambda_1^+(z_2))(z_1 - i(|z_2| + 1))} - \frac{\tilde{\gamma}_2^k(z_2)(-iz_1)^k [i(|z_2| + 1) - \lambda_1^-(z_2)]}{a^{20}(z_1 - \lambda_1^+(z_2))(z_1 - i(|z_2| + 1))^2}$$

$$- \frac{\tilde{\gamma}_2^k(z_2)(-iz_1)^k}{a^{20}(z_1 - \lambda_1^+(z_2))(z_1 - i(|z_2| + 1))} \mathcal{O}\left(\left| \frac{i(|z_2| + 1) - \lambda_1^-(z_2)}{z_1 - i(|z_2| + 1)} \right|^2 \right), \qquad z \in \mathbb{R}^2. \tag{11.4}$$

The first and second terms on the right-hand side of (11.3) are analytic in $\mathrm{Im}\, z_1 < 0$. Hence, their inverse Fourier transform vanish for $x_1 > 0$ by the Paley-Wiener theorem [115, 122]. Therefore, these terms do not contribute to the norm $\|w\|_{H^s(K)}$. Finally, denoting the last term of (11.4) by $\tilde{w}_3(z)$, we get by (7.18),

$$\|w_3\|^2_{H^s(K)} \le \int (|z_1| + |z_2| + 1)^{2s} |\tilde{w}_3(z)|^2 dz$$

$$\le C \int [\int (|z_1| + |z_2| + 1)^{2s+2k-8} dz_1] |\tilde{\gamma}_2^k(z_2)|^2 (|z_2| + 1)^4 dz_2$$

$$\le C_1 \int |\tilde{\gamma}_2^k(z_2)|^2 (|z_2| + 1)^{2s+2k-3} dz_2 = C_1 \|\gamma_2^k\|^2_{H^{s+k-3/2}(\mathbb{R})}.$$

$$\tag{11.5}$$

At last, (8.11) implies that

$$\|\gamma_2^k\|_{H^{s+k-3/2}(\mathbb{R})} \le C \sum_j \|v_2^{j-k-1}\|_{\mathring{H}^{s+k-3/2+2-j}(\mathbb{R}+)} = C \sum_\beta \|v_2^\beta\|_{\mathring{H}^{s-\beta-1/2}(\mathbb{R}+)}.$$

$$\tag{11.6}$$

This proof is valid for $s \in (3/2, 5/2)$, since then $2s+2k-8 < -1$ for $k = 0, 1$. In the case $s \ge 5/2$ the expansion (11.3) should be continued further.

(ii) This follows from Lemma 8.4.   □

Chapter 12
Undetermined Algebraic Equations on the Riemann Surface

We can rewrite (9.9), (10.7) as an algebraic system on the Riemann surface V. By Lemma 7.1 the sets

$$V_l^{\pm} := \{z \in V : \pm \text{Im} \, z_l > 0\}, \quad l = 1, 2. \tag{12.1}$$

are nonempty. Obviously, $V^* := V \cap \mathbb{C}K^* = V_1^+ \cap V_2^+$. Consider the functions

$$v_{l\beta}(z) := \tilde{v}_l^{\beta}(z_l), \quad z \in V_l^+, \quad l = 1, 2, \quad \beta = 0, 1. \tag{12.2}$$

These functions are holomorphic in V_l^+. Now (9.9), (10.7) can be written as a system of three algebraic equations on V^* with four unknown functions:

$$\begin{cases} \displaystyle\sum_{\beta=0,1} A_1^{\beta}(z) v_{1\beta}(z) + \sum_{\beta=0,1} A_2^{\beta}(z) v_{2\beta}(z) = 0, \, z \in V^* \\[2mm] \tilde{B}_{10}(z_1) v_{10}(z) + \tilde{B}_{11}(z_1) v_{11}(z) = g_1(z_1), \quad z \in V_1^+ \\[2mm] \tilde{B}_{20}(z_2) v_{20}(z) + \tilde{B}_{21}(z_2) v_{21}(z) = g_2(z_2), \quad z \in V_2^+ \end{cases} . \tag{12.3}$$

Lemma 9.1 (ii) suggests four combinations. We assume below for definiteness that

$$\deg \tilde{B}_{l0}(z_l) = m_l, \quad z \in V, \quad l = 1, 2, \tag{12.4}$$

which holds, for example, for the Dirichlet b.c. In such case we can eliminate two Dirichlet data v_{l0} from system (12.3) expressing them from the last two equations:

$$v_{l0}(z) = \frac{g_l(z_l) - \tilde{B}_{l1}(z_l) v_{l1}(z)}{\tilde{B}_{l0}(z_l)}, \quad z \in V_l^+, \quad l = 1, 2. \tag{12.5}$$

A. Komech, A. Merzon, *Stationary Diffraction by Wedges*, Lecture Notes in Mathematics 2249, https://doi.org/10.1007/978-3-030-26699-8_12

Substituting into the first equation of (12.3), we obtain one equation with two unknown functions,

$$S_1(z)v_{11}(z) + S_2(z)v_{21}(z) = F(z), \qquad z \in V^*, \tag{12.6}$$

where the coefficients are the polynomials

$$\left. \begin{aligned} S_1(z) &= \tilde{B}_{20}(z_2)\left[-A_1^0(z)\tilde{B}_{11}(z_1) + A_1^1(z)\tilde{B}_{10}(z_1)\right] \\ S_2(z) &= \tilde{B}_{10}(z_1)\left[-A_2^0(z)\tilde{B}_{21}(z_2) + A_2^1(z)\tilde{B}_{20}(z_2)\right] \end{aligned} \right|, \qquad z \in \mathbb{C}^2. \tag{12.7}$$

The right-hand side of (12.6) is an analytic function in $\mathbb{C}K^*$,

$$F(z) = -A_1^0(z)\tilde{B}_{20}(z_2)g_1(z_1) - A_2^0(z)\tilde{B}_{10}(z_1)g_2(z_2), \qquad z \in \mathbb{C}K^*. \tag{12.8}$$

The following proposition is of crucial importance for our method.

Proposition 12.1 *Let A be strongly elliptic operator and let the Shapiro-Lopatinski condition hold as well as (12.4). Then for $l = 1, 2$*

$$|S_l(z)| \geq \varkappa |z|^{m_1+m_2}, \qquad |z| > R, \quad z \in V, \tag{12.9}$$

where $\varkappa > 0$ and $R > 0$ is sufficiently large.   □

We prove this proposition in Chap. 17. The proof relies on the formulas

$$\left. \begin{aligned} S_1(z) &= \tilde{B}_{20}(z_2)\left[-\tilde{B}_{11}(z_1)[a^{02}(-iz_2) + a^{11}(-iz_1) + a^{01}] + \tilde{B}_{10}(z_1)a^{02}\right] \\ S_2(z) &= \tilde{B}_{10}(z_1)\left[-\tilde{B}_{21}(z_2)[a^{20}(-iz_1) + a^{11}(-iz_2) + a^{10}] + \tilde{B}_{20}(z_2)a^{20}\right] \end{aligned} \right|, \qquad z \in \mathbb{C}^2, \tag{12.10}$$

which follow from (12.7) and (10.10).

Remark 12.2 Estimates (12.9) hold for the coefficients of Eq. (12.6) under conditions (12.4). For other combinations of conditions (9.2), the coefficients of the corresponding equation of type (12.6) satisfy an estimate similar to (12.9). Moreover, all subsequent constructions of our method admit suitable extension to these cases.

Chapter 13
Automorphic Functions on the Riemann Surface

The algebraic system (12.3) and Eq. (12.6) are both undetermined. Hence, we need additional equations for the Cauchy data $v_{l\beta}$. We obtain the additional equations by the method of automorphic functions, which was introduced by Malyshev in [77] in the context of difference equations in the quadrant of the plane.

The main idea of Malyshev's method is to formalize the fact that the functions $v_{l\beta}(z)$ depend in fact only on one variable z_l. The suitable formalization is provided by the introduction of the corresponding covering monodromy automorphisms on V.

Definition 13.1 $p_l : V \to \mathbb{C}$ with $l = 1, 2$ is the coordinate projection $p_l : (z_1, z_2) \mapsto z_l$. $\square$

Using the notations (8.7) we can rewrite the equation of the Riemann surface V as

$$V = \{z \in \mathbb{C}^2 : \sum_{j \le 2} \tilde{A}_{1j}(z_1)(-iz_2)^j = 0\} = \{z \in \mathbb{C}^2 : \sum_{j \le 2} \tilde{A}_{2j}(z_2)(-iz_1)^j = 0\}.$$

These equations, which determine the surface V, imply that each projection p_l is a two-sheeted holomorphic *covering map* $p_l : V \to \mathbb{C}$, and p_l^{-1} has two branching points since the symbol $\tilde{A}(z)$ is irreducible. Hence, the equation $z_l = p_l z$ with a fixed z_l admits two roots $z', z'' \in V$, and the roots are distinct away from the branching points.

Definition 13.2 The monodromy automorphism $h_l : V \to V$ with $l = 1, 2$ is defined as

$$h_l(z') = z'' \quad \text{and} \quad h_l(z'') = z' \tag{13.1}$$

if $p_l z' = p_l z''$ and $z' \ne z''$. Otherwise, $h_l(z') = z'$. $\square$

© Springer Nature Switzerland AG 2019
A. Komech, A. Merzon, *Stationary Diffraction by Wedges*, Lecture Notes
in Mathematics 2249, https://doi.org/10.1007/978-3-030-26699-8_13

In other words, the automorphism h_l transposes the points of the Riemann surface with the same coordinates z_l. Equation (8.7) imply, by the Vieta theorem,

$$h_1 z = (z_1, -i\,\tilde{A}_{11}(z_1)/\tilde{A}_{12} - z_2), \qquad h_2 z = (-i\,\tilde{A}_{21}(z_2)/\tilde{A}_{22} - z_1, z_2), \qquad z \in V.$$
$$\tag{13.2}$$

Introduce the exponential notation $v^h(z) \equiv v(hz)$ for a function v on V and a map $h : V \to V$. Let us note that $h_l V_l^{\pm} = V_l^{\pm}$ for $l = 1, 2$. By the definition (13.1),

$$h_l V_l^{\pm} = V_l^{\pm}, \qquad l = 1, 2. \tag{13.3}$$

Now we can formulate new functional algebraic equations, which express the invariance of the functions $v_{l\beta}(z)$ with respect to h_l:

$$v_{l\beta}^{h_l}(z) = v_{l\beta}(z), \qquad z \in V_l^+, \qquad \beta = 0, 1, \qquad l = 1, 2. \tag{13.4}$$

This is a system of four algebraic equations.

The algebraic equations (12.6), (13.4), and (12.5), were obtained for any distributional solution $u \in S'(K)$ to the boundary problem (7.1), (7.2). This combined system of algebraic equations is equivalent to the boundary problem in the sense of the following proposition. For any $v_{l\beta} \in S'(\overline{\mathbb{R}^+})$ let us define $u \in S'(K)$ by the formula (8.5), where the distribution γ is given by (8.3) with densities (8.11).

Proposition 13.3 *Let A be strongly elliptic operator and let the Shapiro-Lopatinski condition hold as well as (12.4). Then*

(i) *For any solutions $v_{l\beta} \in S'(\overline{\mathbb{R}^+})$ to algebraic equations (12.6), (13.4), and (12.5), the corresponding distribution $u \in S'(K)$ is the solution to boundary problem (7.1), (7.2).*

(ii) *For any solutions $v_{l\beta} \in H^{s-\beta-1/2}(\overline{\mathbb{R}^+})$ with $s > 3/2$ to algebraic equations (12.6), (13.4), and (12.5), the corresponding distribution $u \in H^s(K)$ is the solution to boundary problem (7.1), (7.2).* □

Proof (i) Follows from Lemma 11.1, and (ii) follows from Lemma 11.2. □

Chapter 14
Functional Equation with a Shift

Algebraic equation (12.6) contains two unknown functions $v_{l1}(z)$ on the Riemann surface V^*. Hence, this is an undetermined algebraic problem. However, the identities (13.4) give, in particular, that

$$v_{l1}^{h_l}(z) = v_{l1}(z), \quad z \in V_l^+, \quad l = 1, 2. \tag{14.1}$$

We will show that these identities make the problem (12.6) well-posed.

The coefficients $S_1(z)$ and $S_2(z)$ are analytic functions in V by (12.7). However, the function F generally is analytic only in V^* by (12.8). Hence, we cannot apply the automorphisms h_l directly to Eq. (12.6). On the other hand, it suffices to solve the problem first for the case

$$f_2(x_2) \equiv 0, \tag{14.2}$$

and then for the case $f_1(x_1) \equiv 0$. In the case (14.2) the function $F(z)$ is analytic in V_1^+ and in V_2^+ by (12.8) and (9.9), and moreover,

$$|F(z)| \leq C \frac{(|z| + 1)^M}{|\text{Im } z_1|^N}, \quad z \in V^+ := V_1^+ \cup V_2^+. \tag{14.3}$$

Also,

$$S_l(z) \not\equiv 0, \quad z \in V, \qquad l = 1, 2 \tag{14.4}$$

by (12.9). Hence, identity (12.6) implies that $v_{21}(z)$ admits a meromorphic continuation from V_2^+ to V_1^+, and the same identity holds for this continuation,

$$S_1(z)v_{11}(z) + S_2(z)v_{21}(z) = F(z), \quad z \in V^+. \tag{14.5}$$

© Springer Nature Switzerland AG 2019
A. Komech, A. Merzon, *Stationary Diffraction by Wedges*, Lecture Notes
in Mathematics 2249, https://doi.org/10.1007/978-3-030-26699-8_14

Moreover, (9.8) and (14.4) imply the estimate

$$|v_{21}(z)| \leq C \frac{(|z|+1)^{M_{21}}}{|\mathrm{Im}\, z_1|^{N_{21}}}, \qquad z \in V^+ : \mathrm{Im}\, z_1 \in (0, \varepsilon), \qquad (14.6)$$

where $\varepsilon > 0$.

Corollary 14.1 *Each pole of $v_{21}(z)$ in V_1^+ is a zero point $\zeta_j \in V_1^+ \setminus V_2^+$ of S_2. The order of the pole does not exceed the multiplicity m_j of ζ_j:*

$$v_{21}(z) = \sum_{j}^{m_j} \sum_{1} \frac{r_{jk}}{(z - \zeta_j)^k} + f(z), \qquad z \in V_1^+,$$

where $f(z)$ is an analytic function in V_1^+.

Remark 14.2 The meromorphic continuation of $v_{21}(z)$ to $V_1^+ \setminus V_2^+$ is not defined outside the Riemann surface V.

Now we can apply h_1 to (14.5) with $z_1 \in V_1^+$ by (13.3): using (14.1) and (13.3), we get

$$S_1^{h_1}(z)v_{11}(z) + S_2^{h_1}(z)v_{21}^{h_1}(z) = F^{h_1}(z), \quad z \in V_1^+. \qquad (14.7)$$

Finally, eliminating $v_{11}(z)$ from (14.5) and (14.7), we get the following equation for the *meromorphic continuation* of the function v_{21},

$$Q_1(z)v_{21}(z) - Q_2(z)v_{21}^{h_1}(z) = G(z), \quad z \in V_1^+. \qquad (14.8)$$

Here

$$Q_1 = S_1^{h_1} S_2, \qquad Q_2 = S_1 S_2^{h_1}, \qquad G = S_1^{h_1} F - S_1 F^{h_1}. \qquad (14.9)$$

Let us note that polynomials

$$Q_l(z) \not\equiv 0, \qquad z \in V \qquad (14.10)$$

for $l = 1, 2$ by (14.4). Hence, (14.3) implies the estimate

$$|G(z)| \leq C \frac{(|z|+1)^{M_1'}}{|\mathrm{Im}\, z_1|^{N_1}}, \qquad z \in V_1^+. \qquad (14.11)$$

Finally, let us use the invariance of v_{21} with respect to h_2. Namely, (14.1) implies that $v_{21}^{h_1}(z) = v_{21}^{h_2 h_1}(z)$ if $h_1 z \in V_2^+$, i.e., if $z \in h_1 V_2^+$. Hence, denoting $h = h_2 h_1$, we get from (14.8)

$$Q_1(z) v_{21}(z) - Q_2(z) v_{21}^h(z) = G(z), \quad z \in V_1^+ \cap h_1 V_2^+. \tag{14.12}$$

This intersection is nonempty. Indeed, (7.17) implies that this intersection contains, in particular, the points $(z_1 + i\varepsilon, \lambda_2^-(z_1 + i\varepsilon))$ with $z_1 \in \mathbb{R}$ and small $\varepsilon > 0$.

Remark 14.3 We will see that this equation can be reduced to the Riemann-Hilbert problem in contrast to Eq. (14.8), which is the *one-sided Carleman problem* [150]. □

Lemma 14.4 *Let v_{21} be a meromorphic in V^+ solution to (14.12), invariant with respect to h_2 in V_2^+. Then the function v_{11}, as defined by (14.5), is invariant with respect to h_1 in V_1^+.* □

Proof Equation (14.12) implies (14.8) since v_{21} is h_2-invariant. On the other hand, (14.8) and (14.5) imply (14.7). Finally, applying h_1 to (14.5) we obtain (14.7) with $v_{11}^{h_1}$ instead of v_{11}. Hence, $v_{11}^{h_1}$ coincides with v_{11} by (14.4) with $l = 1$. □

Chapter 15
Lifting to the Universal Covering

We are going to reduce Eq. (14.12) to the Riemann-Hilbert problem. For this purpose, we should make the domains V_l^+ and the covering automorphisms h_l clearly visible in appropriate coordinates on the Riemann surface V. We will use the global coordinates on the universal covering surface $\hat{V}$ of V, which give the *uniformization* of the Riemann surface. These coordinates were introduced in [52] for general strongly elliptic second-order operators (7.1).

For definiteness, we will consider until the end of Chap. 19 the strongly elliptic Helmholtz operator $A(\partial)$ defined by (7.4) with $\omega \in i\mathbb{R}^+$. The extension of all further steps to $\omega \in \mathbb{C} \setminus \mathbb{R}$ and to general strongly elliptic operators A is straightforward, see Remark 19.9. For the operator (7.4) we have

$$\tilde{A}_{11}(z) = 2iz_1 \cos \Phi, \quad \tilde{A}_{12}(z) = \tilde{A}_{22} = 1, \quad \tilde{A}_{21} = 2iz_2 \cos \Phi$$

by (8.7). Hence, the covering automorphisms (13.2) read

$$h_1(z_1, z_2) = (z_1, 2z_1 \cos \Phi - z_2), \quad h_2(z_1, z_2) = (2z_2 \cos \Phi - z_1, z_2).$$

The symbol of the operator (7.4) reads $\tilde{A}(z) = -z_1^2 + 2\cos \Phi z_1 z_2 - z_2^2 + \omega^2 \sin^2 \Phi$. Hence the equation of the Riemann surface (10.3) can be written as

$$(z_1 \sin \Phi)^2 + (z_2 - z_1 \cos \Phi)^2 = (\omega \sin \Phi)^2.$$

Now the uniformization is obvious: $z_1 = \omega \sin \varphi$ and $z_2 - z_1 \cos \Phi = \omega \sin \Phi \cos \varphi$, where $\varphi \in \mathbb{C}$. Therefore, $z_2 = \omega(\sin \varphi \cos \Phi + \sin \Phi \cos \varphi) = \omega \sin(\varphi + \Phi)$. We prefer the coordinate $w = i\varphi$, hence finally, we will use the uniformization

$$z_1 = z_1(w) := -i\omega \sinh w, \quad z_2 = z_2(w) := -i\omega \sinh(w + i\Phi), \quad w \in \mathbb{C}. \tag{15.1}$$

© Springer Nature Switzerland AG 2019
A. Komech, A. Merzon, *Stationary Diffraction by Wedges*, Lecture Notes in Mathematics 2249, https://doi.org/10.1007/978-3-030-26699-8_15

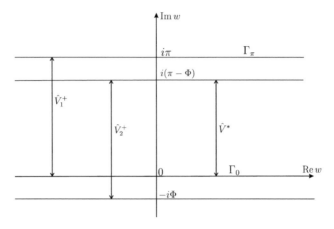

Fig. 15.1 Universal covering in the case $\omega \in i\mathbb{R}^+$

Hence, the surface V is isomorphic to $\mathbb{C} \setminus 0$, and the universal covering $\hat{V}$ is isomorphic to $\mathbb{C}$.

Let us calculate the regions V_l^+ and the monodromy automorphisms h_l in the coordinate w. The formulas (15.1) give the biholomorphism regions V_1^+ and V_2^+ with the regions $\hat{V}_1^+$ and $\hat{V}_2^+$ on $\hat{V}$, where we choose

$$\hat{V}_1^+ = \{w \in \mathbb{C} : 0 < \operatorname{Im} w < \pi\}, \qquad \hat{V}_2^+ = \{w \in \mathbb{C} : -\Phi < \operatorname{Im} w < \pi - \Phi\}. \tag{15.2}$$

Respectively, $V^* = V_1^+ \cap V_2^+$ and $V^+ = V_1^+ \cup V_2^+$ are identified with the covering regions

$$\hat{V}^* = \hat{V}_1^+ \cap \hat{V}_2^+ = \{w : 0 < \operatorname{Im} w < \pi - \Phi\}, \qquad \hat{V}^+ = \hat{V}_1^+ \cup \hat{V}_2^+ = \{w : -\Phi < \operatorname{Im} w < \pi\},$$

see Fig. 15.1.

Remark 15.1

(i) If $\omega \in \mathbb{C}^+$ and $\operatorname{Re} \omega \neq 0$, the operator A remains strongly elliptic and the regions $\hat{V}_l^+$ in this case are similar to the corresponding regions for $\omega \in i\mathbb{R} \setminus 0$ (cf. Figs. 15.1 and 15.2). All subsequent steps of our method allow suitable extensions to this setting.

(ii) If $\omega \in R \setminus 0$, then A is an elliptic operator, but it is not strongly elliptic. In this case the regions V_l^+ become disconnected and the corresponding Riemann–Hilbert problem becomes ill-posed [88].   □

Note that the lifting $\hat{h}_l : \hat{V} \to \hat{V}$ of the generator h_l to $\hat{V}$, satisfying the identity

$$z(\hat{h}_l w) = h_l z(w), \qquad w \in \hat{V}, \tag{15.3}$$

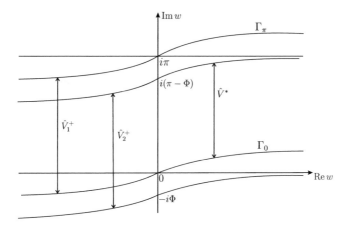

Fig. 15.2   Universal covering in the case $\operatorname{Re} \omega \neq 0$

is non-unique. We choose the branch of $\hat{h}_l$ which maps $\hat{V}_l^+$ onto itself for $l = 1, 2$. Then (15.1) implies that the automorphisms $\hat{h}_l$ act by

$$\begin{cases} \hat{h}_1(w) = -w + i\pi, & w \in \hat{V}_1^+ \\ \hat{h}_2(w) = -w + i\pi - 2i\Phi, & w \in \hat{V}_2^+ \end{cases}. \qquad (15.4)$$

Thus, the automorphism $\hat{h}_1$ is the point reflection with the center $i\pi/2$, and $\hat{h}_2$ is the point reflection with the center $i\pi/2 - i\Phi$.

Finally, let us lift Eq. (14.12) to $\hat{V}$. Namely, denote by $\hat{v}_{l1}(w)$ the lifting of the holomorphic functions v_{l1} from V_l^+ to $\hat{V}_l^+$, and by $\hat{Q}_l$, the lifting of the polynomials Q_l to $\hat{V}$ by formulas (15.1):

$$\hat{v}_{l1}(w) = v_{l1}(z(w)), \quad w \in \hat{V}_l^+; \qquad \hat{G}(w) = G(z(w)), \quad w \in \hat{V}_1^+, \qquad (15.5)$$

$$\hat{Q}_l(w) = Q_l(z(w)), \qquad w \in \hat{V}.$$

By (14.1), (15.3) and (15.5), the functions $\hat{v}_{l1}$ are invariant with respect to $\hat{h}_l$:

$$\hat{v}_{l1}^{\hat{h}_l}(w) = \hat{v}_{l1}(w), \qquad w \in \hat{V}_l^+. \qquad (15.6)$$

Let us denote

$$\Pi_a^b := \{w \in \mathbb{C} : a < \operatorname{Im} w < b\}.$$

Then

$$\hat{V}_1^+ = \Pi_0^\pi, \qquad \hat{V}_2^+ = \Pi_{-\Phi}^{\pi-\Phi}$$

by (15.2), and

$$\hat{h}_1 \hat{V}_2^+ = \Pi_\Phi^{\pi + \Phi}, \qquad \hat{V}_1^+ \cap \hat{h}_1 \hat{V}_2^+ = \Pi_\Phi^\pi$$

by (15.4). Moreover, $\hat{Q}_l$ is a holomorphic $2\pi i$-periodic function of $w \in \mathbb{C}$, and

$$\hat{h}w := \hat{h}_2 \hat{h}_1 w = w - 2i\Phi, \qquad w \in \Pi_\Phi^\pi \tag{15.7}$$

by (15.4). Hence, (14.12) turns into the following difference equation,

$$\hat{Q}_1(w)\hat{v}_{21}(w) - \hat{Q}_2(w)\hat{v}_{21}(w - 2i\Phi) = \hat{G}(w), \qquad w \in \Pi_\Phi^\pi. \tag{15.8}$$

Now Corollary 14.1 implies that

$$\hat{v}_{21}(w) = \sum_j \sum_1^{m_j} \frac{r'_{jk}}{(w - w_j)^k} + g(w), \qquad w \in \Pi_0^\pi = \hat{V}_1^+, \tag{15.9}$$

where $w_j = w(\zeta_j) \in \Pi_{\pi - \Phi}^\pi = \hat{V}_1^+ \setminus \hat{V}_2^+$, while $g(w)$ is an analytic function in Π_0^π.
Let us note that

$$|z| \sim e^{|w|}, \qquad |\operatorname{Im} z_1| \sim e^{|w|} |\operatorname{Im} w - \pi|, \qquad w \in \Pi_{\pi - \Phi}^\pi.$$

Hence, the bounds (14.6) and (14.11) give

$$|\hat{v}_{21}(w)| + |\hat{G}(w)| \le C \frac{(e^{|w|} + 1)^{M'}}{|\operatorname{Im} w - \pi|^{N'}}, \qquad w \in \Pi_{\pi - \varepsilon'}^\pi, \tag{15.10}$$

for small $\varepsilon' > 0$.

Lemma 15.2 *Any meromorphic in $\Pi_{-\Phi}^\pi$ solution $\hat{v}_{21}$ to Eq. (15.8) admits a mero-morphic continuation to $\Pi_{-2\Phi}^{-\Phi}$, satisfying the difference equation*

$$\hat{Q}_1(w)\hat{v}_{21}(w) - \hat{Q}_2(w)\hat{v}_{21}(w - 2i\Phi) = \hat{G}(w), \quad w \in \hat{V}_1^+ = \Pi_0^\pi. \tag{15.11}$$

Proof This assertion follows from (15.8), since $\hat{Q}_2(w) \not\equiv 0$ by (14.10) and since the function $\hat{G}(w)$ is analytic in $\hat{V}_1^+$. $\square$

In the next lemma we prove that system (15.11), (15.6) is equivalent to system (14.5), (14.1).

Lemma 15.3

(i) *Let $v_{l2}(z_l) \in H(V_l)^+$ satisfy (14.5) and (14.1) with $l = 2$. Then $\hat{v}_{21}(w) := v_{21}(z(w))$ satisfies (15.11) and (15.6) with $l = 2$.*

(ii) Conversely, let $\hat{v}_{21}$ satisfy (15.11) and (15.6) with $l = 2$, and let us define

$$\hat{v}_{11}(w) := \frac{F(z(w)) - \hat{S}_2(w)\hat{v}_{21}(w)}{\hat{S}_1(w)}, \qquad w \in \hat{V}_1^+. \qquad (15.12)$$

Then $v_{l1}(z) := \hat{v}_{l1}(z(w))$ with $l = 1, 2$ satisfy (14.5), (14.1).

Proof

(i) The assertion follows from Lemmas 14.4 and 15.2.
(ii) Obviously, $v_{11}(z)$, $v_{21}(z)$ satisfy (14.5) by (15.12), and v_{21} is h_2—invariant
by (15.6) with $l = 2$. Finally, applying the automorphism $\hat{h}_1$ to both parts
of (15.12) we see that the identity $\hat{v}_{11}^{\hat{h}_1} = \hat{v}_{11}$ is equivalent to (15.11) by (14.9).
Hence, v_{11} is h_1-invariant.

$\square$

This lemma means that it suffices to find all $\hat{h}_2$-invariant solutions to the
difference equation (15.11). This $\hat{h}_2$-invariance can be provided by symmetrization.

Lemma 15.4 *Let the function $\hat{v}_{21}(w)$ be meromorphic in $\Pi_{-2\Phi}^{\pi}$ and let it satisfy (15.11). Then its symmetrization*

$$(\hat{v}_{21} + \hat{v}_{21}^{\hat{h}_2})/2 \qquad (15.13)$$

is meromorphic in $\Pi_{-2\Phi}^{\pi}$, satisfies (15.11), and is $\hat{h}_2$-invariant.

Proof The function (15.13) is defined in $\Pi_{-2\Phi}^{\pi}$, since this domain is $\hat{h}_2$-invariant
by its reflection symmetry with the center $i\pi/2 - i\Phi$. It remains to check that $\hat{v}_{21}^{\hat{h}_2}$
satisfies (15.11). Let us write (15.11) as

$$\hat{Q}_1\hat{v}_{21} - \hat{Q}_2\hat{v}_{21}^{\hat{h}} = \hat{G} \qquad (15.14)$$

according to (15.7). Here $\hat{Q}_1 = \hat{Q}_2^{\hat{h}_1}$ by (14.9). Hence, substituting $\hat{v}_{21}^{\hat{h}_2}$ instead of
$\hat{v}_{21}$ into the left-hand side, we obtain

$$\hat{Q}_2^{\hat{h}_1}\hat{v}_{21}^{\hat{h}_2} - \hat{Q}_1^{\hat{h}_1}\hat{v}_{21}^{\hat{h}_1} = [\hat{Q}_2\hat{v}_{21}^{\hat{h}} - \hat{Q}_1\hat{v}_{21}]^{\hat{h}_1} = -\hat{G}^{\hat{h}_1}.$$

by (15.14). Finally, $\hat{G}^{\hat{h}_1} = -\hat{G}$ by (14.9). Hence, Eq. (15.14) for $\hat{v}_{21}^{\hat{h}_2}$ is proven. $\square$

Chapter 16
The Riemann-Hilbert Problem on the Riemann Surface

Here we reduce Eq. (15.11) to the Riemann-Hilbert problem. Let us recall that the function $\hat{v}_{21}$ is meromorphic in $\Pi^\pi_{-2\Phi}$ by (15.9) and Lemma 15.2. Consider the strip

$$W := \Pi^\pi_{\pi-2\Phi},$$

see Fig. 16.1.

Definition 16.1

(i) $\mathcal{M}_1$ is the set of meromorphic functions $\hat{v}_{21}$ in $\Pi^\pi_{-2\Phi}$ which are holomorphic in $V_2^+ = \hat{\Pi}^{\pi-\Phi}_{-\Phi}$ and satisfy the estimate (15.10).

(ii) $\mathcal{M}_2$ is the set of meromorphic functions $\hat{v}_{21}$ in the strip W, which satisfy the estimate of type (15.10),

$$|\hat{v}_{21}(w)| \leq C \frac{(e^{|w|} + 1)^{M'}}{[\text{dist}\,(w, \partial W)]^{N'}}, \qquad \text{dist}\,(w, \partial W) < \varepsilon, \qquad (16.1)$$

where $\varepsilon > 0$.

$\square$

The next lemma shows that Eq. (15.11) is essentially equivalent to the equation

$$\hat{Q}_1(w)\hat{v}_{21}(w - i0) - \hat{Q}_2(w)\hat{v}_{21}(w - 2i\Phi + i0) = \hat{G}(w - i0), \qquad w \in \Gamma_\pi. \qquad (16.2)$$

Here the limit values of all functions exist in the sense of distributions due to the bound (16.1).

© Springer Nature Switzerland AG 2019
A. Komech, A. Merzon, *Stationary Diffraction by Wedges*, Lecture Notes in Mathematics 2249, https://doi.org/10.1007/978-3-030-26699-8_16

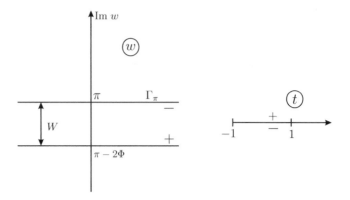

Fig. 16.1 Factorization

Definition 16.2

(i) $\mathcal{E}_1$ is the space of $\hat{v}_{21} \in \mathcal{M}_1$ which are solutions to (15.11).
(ii) $\mathcal{E}_2$ is the space of $\hat{v}_{21} \in \mathcal{M}_2$ which are solutions to (16.2).   □

We will denote

$$\Gamma_a := \{w \in \mathbb{C} : \operatorname{Im} w = a\}.$$

Lemma 16.3

(i) *For any function $\hat{v}_{21} \in \mathcal{E}_1$, its restriction $\hat{v}_{21}|_W \in \mathcal{E}_2$.*
(ii) *Any function $\hat{v}_{21} \in \mathcal{E}_2$, admits a unique continuation which is meromorphic in $\Pi^{\pi}_{-2\Phi}$ and satisfies (15.11).*   □

Proof

(i) Let $\hat{v}_{21}(w) \in \mathcal{M}_1$ be a solution to (15.11). Then

$$\hat{Q}_1(w)\hat{v}_{21}(w - i0) - \hat{Q}_2(w)\hat{v}_{21}(w - 2i\Phi - i0) = \hat{G}(w - i0), \quad w \in \Gamma_{\pi}.$$

However,

$$\hat{v}_{21}(w - 2i\Phi - i0) = \hat{v}_{21}(w - 2i\Phi + i0), \qquad w \in \Gamma_{\pi},$$

since the points $w - 2i\Phi$ lie in the strip $\hat{\Pi}^{\pi - \Phi}_{-\Phi}$, where $\hat{v}_{21}$ is holomorphic. Hence, (15.11) implies (16.2).

(ii) Conversely, let $\hat{v}_{21} \in \mathcal{E}_2$. Let us define the meromorphic function $\hat{v}_{21}(w)$ in the strip $W - 2i\Phi$ via Eq. (15.11):

$$\hat{v}_{21}(w) = \frac{\hat{Q}_1(w + 2i\Phi)\hat{v}_{21}(w + 2i\Phi) - \hat{G}(w + 2i\Phi)}{\hat{Q}_2(w + 2i\Phi)}, \qquad w \in W - 2i\Phi.$$

$$(16.3)$$

Let us note that the piece-wise meromorphic function $\hat{v}_{21}$ obtained in this way is meromorphic on the line $\Gamma_{\pi-2\Phi}$, which is the common boundary of the strips W and $W - 2i\Phi$. Denote by $\mathcal{Z}$ the set of all zero points of $\hat{Q}_2(w + 2i\Phi)$ on $\Gamma_{\pi-2\Phi}$. Then (16.3) implies that in the sense of distributions

$$\hat{v}_{21}(w-i0) = \frac{\hat{Q}_1(w - i0 + 2i\Phi)\hat{v}_{21}(w - i0 + 2i\Phi) - \hat{G}(w - i0 + 2i\Phi)}{\hat{Q}_2(w - i0 + 2i\Phi)}, \qquad w \in \Gamma_{\pi-2\Phi} \setminus \mathcal{Z}.$$

On the other hand, (16.2) implies that

$$\hat{v}_{21}(w+i0) = \frac{\hat{Q}_1(w - i0 + 2i\Phi)\hat{v}_{21}(w - i0 + 2i\Phi) - \hat{G}(w - i0 + 2i\Phi)}{\hat{Q}_2(w + i0 + 2i\Phi)}, \qquad w \in \Gamma_{\pi-2\Phi} \setminus \mathcal{Z}.$$

We conclude that

$$\hat{v}_{21}(w - i0) = \hat{v}_{21}(w + i0), \qquad w \in \Gamma_{\pi-2\Phi} \setminus \mathcal{Z}.$$

Therefore, $\hat{v}_{21}$ is a meromorphic function on the line $\Gamma_{\pi-2\Phi}$, and hence, it is meromorphic in the strip $\Pi^{\pi}_{\pi-4\Phi}$ of width 4Φ. Finally, this meromorphic function satisfies Eq. (16.3), which implies that $\hat{v}_{21}$ can be extended to the meromorphic function in $\Pi^{\pi}_{-2\Phi}$ satisfying (15.11).   □

Let us denote $w_- := w - i0$ and $w_+ := w - 2i\Phi + i0$ for $w \in \Gamma_{\pi}$. Then we can rewrite (16.2) as

$$\hat{Q}_1(w)\hat{v}_{21}(w_-) - \hat{Q}_2(w)\hat{v}_{21}(w_+) = \hat{G}(w_-), \quad w \in \Gamma_{\pi}.$$

Let us transform the strip W conformally onto $\mathbb{C}^* \setminus \overline{I}$, where $\mathbb{C}^*$ is the Riemann sphere and $I := (-1, 1)$. For example, we can define the conformal map by

$$w \mapsto t = t(w) := \tanh \frac{(w - i\pi)\pi}{2\Phi}. \qquad (16.4)$$

In particular,

$$t(i(\pi - \Phi)) = \infty, \qquad t(\pm\infty) = \pm 1. \qquad (16.5)$$

Now (16.2) becomes the Riemann-Hilbert problem for meromorphic function $\check{v}_2^1(t)$ in $\mathbb{C}^* \setminus \overline{I}$,

$$\check{Q}_2^-(t)\check{v}_{21}^+(t) - \check{Q}_1^-(t)\check{v}_{21}^-(t) = -\check{G}^-(t), \qquad t \in I. \tag{16.6}$$

Here we denote

$$\check{v}_{21}(t) := \hat{v}_{21}(w), \quad \check{Q}_l(t) := \hat{Q}_l(w), \quad \check{G}(t) := \hat{G}(w), \qquad t \in \mathbb{C} \setminus \overline{I},$$

and for every function f on $C \setminus \overline{I}$ we set

$$f^\pm(t) := f(t \pm i0), \qquad t \in I.$$

By Lemma 16.3 we should construct all solutions $\check{v}_{21}$ to (16.6) such that the corresponding functions $\hat{v}_{21} \in \mathcal{E}_2$. Moreover, (15.9) implies that

$$\check{v}_{21}(t) = \sum_j^{m_j} \sum_1 \frac{r_{jk}''}{(t - t_j)^k} + h(t), \qquad t \in \mathbb{C} \setminus \overline{I}, \tag{16.7}$$

where $t_j = t(w_j) \in \mathbb{C} \setminus R$, while $h(t)$ is an analytic function in $\mathbb{C} \setminus \overline{I}$.

Let us recall the classical scheme of solutions of the Riemann-Hilbert problem. It consists of two main steps.

I. **Factorization**. First, one should solve the corresponding factorization problem

$$\frac{T^+(t)}{T^-(t)} = R(t), \qquad t \in I, \tag{16.8}$$

where T is a holomorphic function in $\mathbb{C} \setminus \overline{I}$, and

$$R(t) := \frac{\check{Q}_2^-(t)}{\check{Q}_1^-(t)} = \frac{\check{Q}_2(t - i0)}{\check{Q}_1(t - i0)} = \frac{\hat{Q}_2(w)}{\hat{Q}_1(w)}, \qquad t \in I. \tag{16.9}$$

II. **Saltus problem**. Second, the nonhomogeneous equation (16.6) reduces to the "saltus problem"

$$T^+(t)\check{v}_{21}^+(t) - T^-(t)\check{v}_{21}^-(t) = -\frac{T^-(t)\check{G}^-(t)}{\check{Q}_1^-(t)}, \qquad t \in I,$$

where $\check{v}_{21}(t)$ is a meromorphic function in $\mathbb{C}^* \setminus \overline{I}$ with poles and singularities prescribed by Corollary 14.1.

The holomorphic solutions to these problems are given *formally* by the Cauchy type integrals

$$\log T(t) = \frac{1}{2\pi i} \int_{-1}^{1} \frac{\log R(s)}{s-t} ds, \quad T(t)\check{v}_{21}(t) = -\frac{1}{2\pi i} \int_{-1}^{1} \frac{T^-(s)\check{G}^-(s)}{\check{Q}_1^-(s)(s-t)} ds, \quad t \in \mathbb{C} \setminus \bar{I}.$$

(16.10)

However, we should modify the holomorphic function $\check{v}_{21}(t)$ to provide the prescribed poles and singularities.

Remark 16.4 Malyuzhinets reduced the boundary problem in the angle with the impedance b.c. to the Riemann-Hilbert problem using slightly different approach [4, 76]. In that context the corresponding factorisation $T(t)$ is called a "Malyuzhinets function".   ☐

Chapter 17
The Factorization

To study properties of the factorization (16.8), we need some preparatory steps. First, we start with the proof of Proposition 12.1. Then we study the edge-point values. Finally, we will reduce the problem to the one with equal edge-point values.

17.1 Ellipticity and Bound Below

Let us prove Proposition 12.1. For definiteness, we will prove this proposition for the operator (7.4) with $\omega \in i\mathbb{R}^+$. Formulas (12.10) can be rewritten as

$$S_1(z) = \tilde{B}_{20}(z_2)a^{02}\left[\tilde{B}_{11}(z_1)(-i[-z_2 - i\frac{a^{11}(-iz_1) + a^{01}}{a^{02}}]) + \tilde{B}_{10}(z_1)\right] = a^{02}\tilde{B}_{20}(z_2)\tilde{B}_1(h_1z)\Bigg|$$

$$S_2(z) = \tilde{B}_{10}(z_1)a^{20}\left[\tilde{B}_{21}(z_2)(-i[-z_1 - i\frac{a^{11}(-iz_2) + a^{10}}{a^{20}}]) + \tilde{B}_{20}(z_2)\right] = a^{20}\tilde{B}_{10}(z_1)\tilde{B}_2(h_2z)\Bigg|$$

due to (13.2) and (9.7). By (12.4) and (7.18) it suffices to check that

$$|\tilde{B}_l(h_lz)| \geq \varkappa|z|^{m_l}, \qquad |z| > R, \quad z \in V \qquad (17.1)$$

with $\varkappa > 0$ for $l = 1, 2$ and large $R > 0$.

For example, we consider the case $l = 1$. Let us set $\lambda := e^w$. Then for $z \in V$

$$z_1 \sim \lambda - \lambda^{-1}, \qquad z_2 \sim c\lambda - (c\lambda)^{-1}, \qquad \lambda^{\hat{h}_1} = -\lambda^{-1}, \qquad \lambda^{\hat{h}_2} = -(c^2\lambda)^{-1} \qquad (17.2)$$

by (15.1) and (15.4), where $c := e^{i\Phi}$. These formulas imply that

$$h_1z \sim (\lambda - 1/\lambda, -c/\lambda + \lambda/c), \qquad z \in V.$$

© Springer Nature Switzerland AG 2019
A. Komech, A. Merzon, *Stationary Diffraction by Wedges*, Lecture Notes in Mathematics 2249, https://doi.org/10.1007/978-3-030-26699-8_17

Therefore,

$$\tilde{B}_1(h_1 z) = \sum_{|j| \le m_1} b_j \lambda^j, \qquad z \in V.$$

Hence, the Shapiro-Lopatinski condition (7.21) and formulas (17.2) imply that the last sum grows like $|\lambda|^{m_1}$ as $|\lambda| \to \infty$ and like $|\lambda|^{-m_1}$ as $|\lambda| \to 0$. Now (17.2) implies (17.1) with $l = 1$. For $l = 2$ the proof is similar. Proposition 12.1 is proved.

17.2   The Edge-Point Values

Lemma 17.1 *There exist limit values* $R(\pm 1) \ne 0$, *and*

$$R(1)R(-1) = 1. \tag{17.3}$$

Proof (16.9) implies that

$$R(t) = \frac{\hat{S}_1^{\hat{h}_1}(w)\hat{S}_2(w)}{\hat{S}_1(w)\hat{S}_2^{\hat{h}_1}(w)}, \qquad t \in I. \tag{17.4}$$

Previous arguments demonstrate that

$$\hat{S}_1(w) = \sum_{|j| \le m_1 + m_2} s_j \lambda^j, \qquad w \in W, \tag{17.5}$$

where the right-hand side is a rational function with poles of the same order $m_1 + m_2$ at $\lambda = 0$ and $\lambda = \infty$. The same is true for all factors in (17.4). Therefore, the quotients $\hat{S}_1(w)/\hat{S}_2(w)$ and $\hat{S}_2(w)/\hat{S}_1(w)$ are well-defined in the limits $w_1 := \operatorname{Re} w \to \mp\infty$ which correspond to $\lambda = 0$ and $\lambda = \infty$. Hence, (17.4) and (16.5) imply that

$$R(-1) := \lim_{t \to -1} R(t) = \lim_{w_1 \to -\infty} \frac{\hat{S}_1(\hat{h}_1 w)}{\hat{S}_2(\hat{h}_1 w)} \frac{\hat{S}_2(w)}{\hat{S}_1(w)} = \lim_{w_1 \to +\infty} \frac{\hat{S}_1(w)}{\hat{S}_2(w)} \lim_{w_1 \to -\infty} \frac{\hat{S}_2(w)}{\hat{S}_1(w)},$$

since $\hat{h}_1$ transposes the points $w = \pm\infty$ by (15.4). Similarly,

$$R(1) := \lim_{t \to 1} R(t) = \lim_{w_1 \to +\infty} \frac{\hat{S}_1(\hat{h}_1 w)}{\hat{S}_2(\hat{h}_1 w)} \frac{\hat{S}_2(w)}{\hat{S}_1(w)} = \lim_{w_1 \to -\infty} \frac{\hat{S}_1(w)}{\hat{S}_2(w)} \lim_{w_1 \to +\infty} \frac{\hat{S}_2(w)}{\hat{S}_1(w)},$$

which implies (17.3).   □

17.3 Equating Edge-Point Values

Let us note that the factorization (16.10) is not optimal, since the Cauchy integral has a logarithmic asymptotics at the endpoints $t = \pm 1$ if $R(\pm 1) \neq 1$. First, we reduce the problem to the unit endpoint values.

Lemma 17.2 *There exists a function $E(t)$ which is analytic and nonzero in $\mathbb{C}^* \setminus \overline{I}$ and such that*

$$\frac{E(t + i0)}{E(t - i0)} \to R(\pm 1) \quad \text{as} \quad t \to \pm 1, \quad t \in I. \tag{17.6}$$

Proof Our construction relies on the identity (17.3). Namely, we define this function as

$$E(t) = [q(t)]^{\log R(1)}, \qquad t \in \mathbb{C}^* \setminus \overline{I}, \tag{17.7}$$

where $q(t)$ is analytic and nonzero in $\mathbb{C}^* \setminus \overline{I}$, and

$$\frac{q(t + i0)}{q(t - i0)} \to e^{\pm 1} \quad \text{as} \quad t \to \pm 1, \quad t \in I; \qquad \text{Im} \log R(1) \in (0, 2\pi). \tag{17.8}$$

We choose

$$q(t) := \left(\frac{t^2 - 1}{\mathfrak{z}^2(t)}\right)^{\frac{1}{2\pi i}}, \qquad t \in \mathbb{C}^* \setminus \overline{I}, \tag{17.9}$$

where $\mathfrak{z}(t)$ is the Zhukovsky function

$$t = \frac{1}{2}\left(\mathfrak{z}(t) + \frac{1}{\mathfrak{z}(t)}\right), \qquad t \in \mathbb{C} \setminus \overline{I}$$

with $\mathfrak{z}(\infty) = \infty$. Obviously,

$$\mathfrak{z}(t) = t + \sqrt{t^2 - 1}, \qquad t \in \mathbb{C} \setminus \overline{I},$$

where $\sqrt{X} > 0$ for $X > 0$. Hence,

$$\mathfrak{z}(\pm 1) = \pm 1. \tag{17.10}$$

This function is a biholomorphism of $\mathbb{C} \setminus \overline{I}$ onto the exterior of the unit circle with center at 0. Hence,

$$\operatorname*{Var}_{\Gamma} \arg \mathfrak{z}(t) = 2\pi, \tag{17.11}$$

where Γ is the contour combined by the intervals $I + i0$ and $I - i0$ and surrounding the interval I in the positive (counter-clockwise) direction. On the other hand,

$$\operatorname*{Var}_{\Gamma} \arg(t^2 - 1) = 4\pi. \tag{17.12}$$

These variations imply that the function (17.9) is analytic in $\mathbb{C} \setminus \overline{I}$. The difference between (17.11) and (17.12) is that the variation (17.11) is continuously distributed on Γ while the variation (17.12) is concentrated near the edge points $t = \pm 1$. Indeed, $\arg[(t \pm i0)^2 - 1]$ does not depend on $t \in (-1, +1)$, while

$$\operatorname*{Var}_{\theta \in [-\pi, \pi]} \arg[(1 + i\varepsilon e^{i\theta})^2 - 1] = \operatorname*{Var}_{\theta \in [0, 2\pi]} \arg[(-1 + i\varepsilon e^{i\theta})^2 - 1] = 2\pi$$

for small $\varepsilon > 0$. This proves (17.12) and (17.8).

Let us note that

$$\frac{t^2 - 1}{\mathfrak{z}^2(t)} \sim (t \mp 1), \qquad t \to \pm 1, \qquad t \in \mathbb{C} \setminus \overline{I}$$

by (17.10). Hence, by (17.7)–(17.9)

$$E(t) \sim (t \mp 1)^\mu, \qquad t \to \pm 1, \tag{17.13}$$

where

$$\mu = \frac{\log R(1)}{2\pi i}. \tag{17.14}$$

Hence, (17.6) follows.   □

Now it remains to factorize the function

$$R_1(t) := \frac{E(t - i0)}{E(t + i0)} R(t), \qquad t \in I$$

with the unit endpoint values: (17.6) implies that

$$R_1(t) \to 1, \qquad t \to \pm 1. \tag{17.15}$$

17.4 Unwinding the Symbol

Proposition 17.3 *There exists an analytic function $T(t)$ in $\mathbb{C} \setminus \overline{I}$, and analytic up to I from both sides, satisfying the relation*

$$\frac{T^+(t)}{T^-(t)} = R(t), \qquad t \in I. \tag{17.16}$$

Moreover,

$$T(t) \neq 0, \quad t \in \mathbb{C} \setminus \bar{I}; \qquad T(t) \sim t^N, \quad t \to \infty$$

with an integer $N \in \mathbb{Z}$.

Proof We want to define the function $\log R_1(t)$ for $t \in I$ with zero values at the endpoints. For this purpose, we should "unwind" the symbol $R_1(t)$.

Let us denote by $\mathcal{Z} := \{t \in I : R_1(t) = 0\}$ the set of zeros and by $\mathcal{P} := \{t \in I : R_1(t) = \infty\}$ the set of poles. Let $\tilde{I}$ denote a smooth contour in $\mathbb{C}$ which coincides with I outside a small neighborhood of $\mathcal{Z} \cup \mathcal{P}$, and in addition, $\tilde{I}$ goes around the zero points from above, and around the poles from below. We set

$$N = -\frac{1}{2\pi} \underset{\tilde{I}}{\mathrm{Var\,arg}}\, R_1(t),$$

where $N \in \mathbb{Z}$ by (17.15). On the other hand, (17.11) implies that

$$\underset{\tilde{I}}{\mathrm{Var\,arg}}\, \frac{\mathfrak{z}(t - i0)}{\mathfrak{z}(t + i0)} = 2\pi, \tag{17.17}$$

where the boundary values $\mathfrak{z}(t \pm i0)$ are defined as analytic continuations from the upper and lower complex half-planes respectively (Fig. 17.1).

Hence, the argument of the function

$$R_0(t) := \left(\frac{\mathfrak{z}(t - i0)}{\mathfrak{z}(t + i0)} \right)^N R_1(t)$$

has zero variation along $\tilde{I}$ by (17.11), where the values $R_1(t)$ for $t \in \tilde{I}$ are defined by analytic continuation from the lower complex half-plane in accordance with (16.9). Moreover,

$$R_0(t) \to \pm 1, \qquad t \to \pm 1 \tag{17.18}$$

by (17.15) and (17.10). Hence, $\log R_0(t)$ is a smooth function on $\tilde{I}$, and $\log R_0(\pm 1) = 0$. Finally, we define $\log R_0(t)$ for $t \in I$ choosing the branches

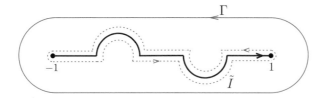

Fig. 17.1 Factorization

according to the bypass rules for $\tilde{I}$. Then this function is analytic above the points of $\mathcal{Z}$ and below the points of $\mathcal{P}$.

Let us define the factorization of $R_0(t)$ from a solution of the "saltus problem"

$$\log T_0^+(t) - \log T_0^-(t) = \log R_0(t), \qquad t \in I.$$

By the Plemelj formulas we can set

$$\log T_0(t) = \frac{1}{2\pi i} \int_I \frac{\log R_0(s)}{s - t} ds, \qquad t \in \mathbb{C}^* \setminus I. \tag{17.19}$$

Then the function $T_0(t)$ is nonzero in $\mathbb{C} \setminus \overline{I}$ and holomorphic in $\mathbb{C} \setminus \overline{I}$ up to I from both sides by the special choice of $\log R_0(t)$. Indeed, for $t_0 \in \mathcal{Z}$,

$$\log T_0^-(t) := \lim_{t \to t_0 - i0} \int_I \frac{\log R_0(s)}{s - t} ds = \int_{\tilde{I}} \frac{\log R_0(s)}{s - t} ds,$$

where the right hand side is an analytic function in a neighborhood of the point t_0. Therefore, the function T_0^- is analytic at the point t_0. Hence, $T_0^+(t) \equiv T_0^-(t) R_0(t)$ is also analytic in the neighborhood of the point t_0. Similar arguments prove the analyticity of $T_0(t)$ at the points $t \in \mathcal{P}$ and for all $t \in I$.

Finally, T_0 does not vanish in $\mathbb{C} \setminus \overline{I}$, and

$$T_0(\infty) = 1.$$

Thus we have

$$\frac{T_0^+(t)}{T_0^-(t)} = R_0(t) = \left(\frac{\mathfrak{z}(t - i0)}{\mathfrak{z}(t + i0)} \right)^N \frac{E(t - i0)}{E(t + i0)} R(t), \qquad t \in I.$$

It remains to define

$$T(t) := \mathfrak{z}^N(t) E(t) T_0(t), \qquad t \in \mathbb{C} \setminus \overline{I}. \tag{17.20}$$

This proves Proposition 17.3.   □

17.5   Asymptotics of the Factorization

We will need the asymptotics of the factorization at the edge points ± 1.

Proposition 17.4 *The function $T(t)$ admits the following asymptotics at the edge points*

$$T(t) \sim T_\pm (t \mp 1)^\mu, \qquad T_\pm \neq 0, \qquad t \to \pm 1, \tag{17.21}$$

where μ is given by (17.14), and Re $\mu \in [0, 1)$ *by (17.8).*   □

Proof Using (17.5) and similar expansions for all factors in (17.4), we obtain that

$$\hat{R}(w) \to R_0^\pm, \qquad \text{Re } w \to \pm\infty, \quad w \in \Gamma_\pi, \qquad (17.22)$$

where $R_0^- R_0^+ = 1$. Further, (16.4) gives

$$e^{\mp w} = (t \mp 1)^v e_\pm(t),$$

where $v = \frac{\pi}{\Phi}$, and functions $e_\pm$ are holomorphic in a neighborhood of ± 1. Now (17.22) implies that

$$\check{R}(t) \to R_0^\pm, \qquad t \to \pm 1.$$

By (17.18)

$$\log R_0(t) \to 0, \qquad t \to \pm 1. \qquad (17.23)$$

Now (17.23) and (17.19) imply the existence of limits

$$\log T_0(t) \to l_\pm, \qquad t \to \pm 1.$$

Hence, exponentiating, we obtain

$$T_0(t) \to T_0^\pm \neq 0, \qquad t \to \pm 1.$$

Now (17.21) follows from (17.20), (17.10), and (17.13).   □

Corollary 17.5 *The function* $\hat{T}(w) := \check{T}(t(w))$ *for* $w \in \Pi^\pi_{-\Phi}$ *admits an extension to a meromorphic function on* $\hat{V}$ *and has a finite number of zeros and poles in* $\Pi^\pi_{-2\Phi}$.

Proof The identity (17.16) is equivalent to

$$\frac{\hat{T}^+(w)}{\hat{T}^-(w)} = \hat{R}(w), \quad w \in \Gamma_\pi. \qquad (17.24)$$

It remains to observe that the right-hand side is a rational function of $\lambda = e^w$.   □

Chapter 18
The Saltus Problem and Final Formula

Here we construct solutions $\check{v}_{21}$ to the Riemann-Hilbert problem (16.6).

18.1 The Saltus Problem

Multiplying Eq. (16.6) by the smooth function $T^-(t)$ we obtain

$$T^+(t)\check{Q}_1^-(t)\check{v}_{21}^+(t) - T^-(t)\check{Q}_1^-(t)\check{v}_{21}^-(t) = -T^-(t)\check{G}^-(t), \qquad t \in I, \qquad (18.1)$$

since $T^-(t)\check{Q}_2^-(t) \equiv T^+(t)\check{Q}_1^-(t)$ by (16.9) and (16.8). Dividing by $\check{Q}_1^-(t)$, we obtain the equivalent "saltus problem"

$$T^+(t)\check{v}_{21}^+(t) - T^-(t)\check{v}_{21}^-(t) = -p.f.\frac{T^-(t)\check{G}^-(t)}{\check{Q}_1^-(t)} + \sum_{\tau \in \mathcal{Z}_1, 0 \leq k < M(\tau)} C_k(\tau)\delta^{(k)}(t-\tau), \qquad t \in I,$$
$$(18.2)$$

where $\mathcal{Z}_1$ is the finite set of zeros $\tau \in I$ of the analytic function Q_1^-, and $p.f.$ means a regularization of the corresponding function on the interval I. The number $M(\tau)$ is the multiplicity of the zero point τ. From the bound (16.1) and (16.4) we obtain

$$|\check{v}_{21}(t)| \leq C\frac{(t^2-1)^{-\Phi M'/\pi}}{[\text{dist}\,(t, I)]^{N'}}, \qquad t \in \mathbb{C} \setminus \overline{I} \qquad (18.3)$$

© Springer Nature Switzerland AG 2019
A. Komech, A. Merzon, *Stationary Diffraction by Wedges*, Lecture Notes in Mathematics 2249, https://doi.org/10.1007/978-3-030-26699-8_18

in a neighborhood of I, where M', N' are given by (16.1). Let us extend the saltus problem to $t \in \mathbb{R}$:

$$[T(t)\check{v}_{21}(t)]\Big|_{t-i0}^{t+i0} = -L[p.f.\frac{T^-(t)\check{G}^-(t)}{\check{Q}_1^-(t)}]$$

$$+ \sum_{\tau \in \mathcal{Z}_1, 0 \le k < M(\tau)} C_k(\tau)\delta^{(k)}(t - \tau) + \sum_{0 \le k \le K} D_k^{\pm}\delta^{(k)}(t \mp 1), \qquad t \in \mathbb{R}.$$

$$(18.4)$$

Here Lf denotes a tempered distribution on $\mathbb{R}$, which is an extension by zero of a distribution f on the open interval I. Such extensions exist for distributions under consideration by the estimates of type (18.3). For example, let us denote the primitive

$$P(t) = \int_i^{t+i0} dt_1 \int_i^{t_1} \dots dt_{K-1} \int_i^{t_{K-1}} dt_K \check{v}_{21}^+(t_K), \qquad t \in \mathbb{R}, \ t_1, t_2, \dots, t_K \in \mathbb{C}^+.$$

$$(18.5)$$

The bound (18.3) implies that this primitive is continuous on $\overline{\mathbb{C}^+}$ if

$$K \ge \max(\overline{M}, \Phi M'/\pi + N') + 1, \qquad \overline{M} := \max(M(\tau) : \tau \in \mathcal{Z}_1).$$

It remains to set

$$L[\check{v}_{21}^+(t)] := \partial_t^K P_0(t) \quad \text{for} \quad t \in \mathbb{R}, \qquad \text{where} \quad P_0(t) := \begin{cases} P(t), \ t \in I \\ 0, \qquad t \in \mathbb{R} \setminus \overline{I} \end{cases}.$$

$$(18.6)$$

Similar definition of Lf applies in (18.4) due to an estimate of type (18.3) for $G(t)$ and the asymptotics (17.21).

18.2   The Final Formula

Let us denote by $\langle \cdot, \cdot \rangle$ duality between distributions with compact support and smooth functions of one variable.

Lemma 18.1

(i) *Let $\hat{v}_{21} \in \mathcal{E}_2$ (see Definition 16.2). Then the corresponding function $\check{v}_{21}$ is given by the "Cauchy type integral"*

$$T(t)\check{v}_{21}(t) = \frac{1}{2\pi i}\langle f(s), \frac{1}{s-t}\rangle + r(t), \qquad t \in \mathbb{C} \setminus \overline{I}, \qquad (18.7)$$

where $f(t)$ is the right hand side of (18.4), and $r(t)$ belongs to a finite-dimensional space of rational functions with poles in $\mathbb{C} \setminus I$.

(ii) *Conversely, let $\check{v}_{21}$ be given by (18.7) with a rational r which has poles only in $\mathbb{C} \setminus I$. Then $\hat{v}_{21} \in \mathcal{E}_2$.*   □

Proof

(i) First, the difference

$$r(t) := T(t)\check{v}_{21}(t) - \frac{1}{2\pi i}\langle f(s), \frac{1}{s-t}\rangle, \qquad t \in \mathbb{C} \setminus \overline{I} \tag{18.8}$$

is analytic for $t \in \mathbb{R}$ by (18.4). Furthermore, $\hat{v}_{21} \in \mathcal{E}_2$, and hence, $\check{v}_{21}$ is meromorphic in $\mathbb{C} \setminus I$ with a finite set of poles and of algebraic growth at infinity by (16.7). Then the product $T(t)\check{v}_{21}(t)$ is a meromorphic function in $\mathbb{C} \setminus I$ with a finite set of poles and of algebraic growth at infinity by Proposition 17.3. Finally, the last term of (18.8) is analytic outside I and of algebraic growth at infinity. Therefore, r is a rational function. It belongs to a finite-dimensional space, since (16.7) specifies the poles and their orders.

(ii) The function (18.7) is meromorphic in $\mathbb{C} \setminus I$ and of an algebraic growth near I and at infinity. Hence, the same properties hold for $\check{v}_{21}$ by Propositions 17.3 and 17.4. In particular, estimate (16.1) holds.

Moreover, (18.2) holds since r is analytic at the points of I. Therefore, (18.4) also holds, which implies (18.1), (16.6) and (16.2). Therefore, $\hat{v}_{21} \in \mathcal{E}_2$.   □

Corollary 18.2 *The function $\check{v}_{21}$ is given by*

$$\check{v}_{21}(t) = \frac{1}{2\pi i\, T(t)}\langle f(s), \frac{1}{s-t}\rangle + \frac{r(t)}{T(t)}, \qquad t \in \mathbb{C} \setminus \overline{I}, \tag{18.9}$$

where $f(t)$ is the right hand side of (18.4), and $r(t)$ belongs to the finite-dimensional space of rational functions of type (16.7).

Chapter 19
The Reconstruction of Solution and the Fredholmness

Now we can reconstruct the solution to the boundary value problem (7.1)–(7.2) in the class of distributional solutions (7.22) and in the class of regular solutions (7.23).

Recall that we consider the strongly elliptic Helmholtz operator (7.4) with $\omega \in i\mathbb{R}^+$, and we assume the Shapiro-Lopatinski condition (7.8). Moreover, we assume the condition (12.4), which can be substituted by different combinations of conditions (9.2) by Lemma 9.1, see Remark 12.2.

19.1   Reconstruction of Distributional Solutions

For distributional boundary data $f_l \in S'(\mathbb{R}^+)$ and solutions $u \in S'(K)$, the Fourier transforms of the extensions $f_l^0 \in S'(\mathbb{R}^+)$ and $u_0 \in S'(\overline{K})$ are characterized by the estimates of type (9.8).

Definition 19.1

(i) $D_1(M, N)$ denotes the linear space of distributions $f_l \in S'(\mathbb{R}^+)$ allowing an extension by zero $f_l^0 \in S'(\mathbb{R})$ satisfying the estimates (9.8).

(ii) $D_2(M, N)$ denotes the linear space of distributions $u \in S'(K)$ allowing an extension by zero $u_0 \in S'(\mathbb{R}^2)$ satisfying the estimates

$$|\tilde{u}_0(z)| \leq C \frac{(|z| + 1)^M}{|\operatorname{Im} z|^N}, \qquad z \in \mathbb{C}K^*. \tag{19.1}$$

$\square$

We will fix a *linear correspondence* between tempered distributions from $D_1(M, N)$ with their extensions by zero for negative argument. For example, such an extension can be defined by the method (18.5), (18.6).

© Springer Nature Switzerland AG 2019
A. Komech, A. Merzon, *Stationary Diffraction by Wedges*, Lecture Notes in Mathematics 2249, https://doi.org/10.1007/978-3-030-26699-8_19

Let us outline the construction of all distributional solutions $u \in D_2(M, N)$ to the boundary value problem (7.1), (7.2) for boundary data $f_l \in D_1(M, N)$ with bounded parameters M, N. It suffices to consider the case $f_2 = 0$. In this case we can define g_2 by (9.9) with $\tilde{f}_2^0 = 0$. The steps of our MAF method are as follows.

I. The solution $\check{v}_{21}$ to the Riemann-Hilbert problem (16.6) is given by formula (18.9) with an arbitrary rational function $r(t)$ of type (16.7). Then the corresponding function $\hat{v}_{21}$ is meromorphic in $W = \Pi_{-2\Phi}^{\pi}$ and satisfies (15.11) by Lemma 16.3 (ii). Then (14.12) holds for the corresponding function v_{21}.

II. The function $\hat{v}_{21}$ should be analytic in the strip $\Pi_{-\Phi}^{\pi-\Phi} = \hat{V}_2^+$. This condition can be written as

$$L_j^1(f_1, C) = 0, \qquad j = 1, \dots, M_1, \tag{19.2}$$

where $C = (C_1, \dots, C_N)$ is the finite-dimensional vector formed by all constants from (9.9), (16.7) and (18.4), and L_j^1 are linear functionals on $D_1(M, N) \oplus \mathbb{C}^N$.

III. Formula (15.13) gives an $\hat{h}_2$-invariant function $\hat{v}_{21}$, which is also a solution to the Riemann-Hilbert problem (16.6), and then (14.12) implies (14.8).

IV. Now we can define v_{11} in V_1^+ from (14.5). It is a meromorphic function by (12.9), and its analyticity in V_1^+ gives additional conditions of type (19.2),

$$L_j^2(f_1, C) = 0, \qquad j = 1, \dots, M_2. \tag{19.3}$$

The invariance of v_{11} with respect to h_1 follows from Lemma 14.4.

V. The functions v_{l0} are defined by (12.5) with an appropriate regularization if $\tilde{B}_{l0}(z_l)$ vanishes at some points $z_l \in \mathbb{R}$. These functions v_{l0} are invariant with respect to h_l since so are v_{l1}. Their analyticity in V_l^+ requires additional conditions

$$L_j^3(f_1, C) = 0, \qquad j = 1, \dots, M_3. \tag{19.4}$$

Finally, the Paley-Wiener theorem implies that all the constructed functions $v_{l\beta}(z_l)$ are the Fourier transforms of tempered distributions $v_l^\beta(x_l) \in S'(\overline{\mathbb{R}^+})$ due to the analyticity of $v_{l\beta}(z_l)$ under the orthogonality conditions (19.2)–(19.4) and due to their estimates.

VI. Now formula (8.5) gives the distributional solution to (7.1), (7.2) by Lemma 11.1.

As a conclusion, we obtain the following theorem.

Theorem 19.2 *Let $A(\partial)$ be the strongly elliptic Helmholtz operator (7.4) with $\omega \in i\mathbb{R}^+$ and let the Shapiro-Lopatinski condition holds. Then in the case (12.4)*

(i) *For any* $K > 0$ *the linear space of solutions* $u \in D_2(M, N)$ *to the homogeneous boundary value problem (7.1), (7.2) with* $M + N \leq K$ *and zero boundary data* $f_1 = f_2 = 0$ *is finite-dimensional.*

(ii) *For any* $K, L > 0$ *a solution* $u \in D_2(M, N)$ *to the problem (7.1), (7.2) with* $M + N \leq K$ *exists for a linear space of boundary data* $f_l \in D_1(M, N)$ *with* $M + N \leq L$ *which satisfy a finite number of orthogonality conditions*

$$L_j(f_1, f_2) = 0, \qquad j = 1, \ldots, J, \qquad (19.5)$$

where L_j *are linear functionals on* $D_1(M, N) \oplus D_1(M, N)$. □

Proof (i) follows from the construction above, since the number of unknown constants $C_1, \ldots, C_N$ is finite, and (ii) holds, since the number of the orthogonality conditions (19.2)–(19.4) is finite. □

19.2   Reconstruction of Regular Solutions

Now let us construct all regular solutions (7.23) to the boundary value problem (7.1), (7.2) with $f_l \in H^{s-m_l-1/2}(\mathbb{R}^+)$, where $s > \max(3/2, m_1 + 1/2, m_2 + 1/2)$ and $s \notin \mathbb{N} + 1/2$. All distributional solutions are constructed above. It remains to select regular solutions $u \in H^s(K)$.

As above, we consider the operator (7.4) with $\omega \in i\mathbb{R}^+$, and we assume the Shapiro-Lopatinski condition (7.8). However, now we strengthen the assumption (12.4) as follows:

$$\tilde{B}_{l0}(z_l) \geq \varkappa(|z_l| + 1)^{m_l}, \qquad z_l \in \mathbb{R}, \quad l = 1, 2, \qquad (19.6)$$

where $\varkappa > 0$ (cf. (17.1)). For example, this inequality holds for the Dirichlet b.c.

I. Again it suffices to consider the case $f_2 = 0$. In this case g_2 is defined by (9.9) with $\tilde{f}_2^0 = 0$. Moreover, it suffices to consider f_1 from the space $\mathring{H}^{s-m_1-1/2}(\mathbb{R}^+)$ of finite codimension in $H^{s-m_1-1/2}(\mathbb{R}^+)$ since $n \notin \mathbb{N} + 1/2$. Then we can take $f_1^0 \in H^{s-m_1-1/2}(\mathbb{R})$. An extension to general case $f_1 \in H^{s-m_1-1/2}(\mathbb{R}^+)$ is a routine matter since $\mathring{H}^{s-m_1-1/2}(\mathbb{R}^+)$ is the subspace of finite codimension in $H^{s-m_1-1/2}(\mathbb{R}^+)$ by Lemma A.1.

II. The distributional Cauchy data $v_l^\beta(x_l) \in S'(\overline{\mathbb{R}^+})$ can be constructed as above under a finite number of orthogonality conditions (19.5). Relation (11.1) implies that it suffices to check that $v_l^\beta \in \mathring{H}^{s-\beta-1/2}(\mathbb{R}^+)$ also under a finite number of orthogonality conditions:

$$\int_{\mathbb{R}} (|z_l| + 1)^{2s-2\beta-1} |\tilde{v}_l^\beta(z_l)|^2 dz_l < \infty. \qquad (19.7)$$

Then the corresponding solution to the problem (7.1), (7.2) belongs to $H^s(K)$ by Lemma 11.2.

III. By (15.1) and (13.4) the condition (19.7) is equivalent to

$$\int_{\Gamma_0 \cup \Gamma_\pi} e^{(2s-2\beta)|w|} |\hat{v}_{1\beta}(w)|^2 dw < \infty, \qquad \beta = 0, 1, \qquad (19.8)$$

together with

$$\int_{\Gamma_{-\phi} \cup \Gamma_{\pi-\phi}} e^{(2s-2\beta)|w|} |\hat{v}_{2\beta}(w)|^2 dw < \infty, \qquad \beta = 0, 1. \qquad (19.9)$$

IV. First, we will prove (19.9) for $\beta = 1$. Let us rewrite the saltus problem (18.2) in the variable $w \in W$ as

$$\hat{T}^+(w)\hat{v}_{21}^+(w) - \hat{T}^-(w)\hat{v}_{21}^-(w) = -p.f. \frac{\hat{T}^-(w')\hat{G}^-(w')}{\hat{Q}_1^-(w')} + \sum_{0 \le k < \hat{M}(v), v \in \hat{Z}_1} \hat{C}_k(v)\delta^{(k)}(w-v)^k, \qquad w \in \Gamma_\pi, \qquad (19.10)$$

where $\hat{Z}_1$ is the finite set of roots of $\hat{Q}_1^-(w)$ on Γ_π, and $\hat{M}(v)$ are the multiplicities of these roots. Now we slightly modify formula (18.7). Namely, (19.10) implies that

$$\hat{T}(w)\hat{v}_{21}(w) = \frac{1}{2\pi i} \langle \hat{g}(w), \frac{e^{-(w-w')^2}}{w-w'} \rangle_{\Gamma_\pi} + \hat{r}(w), \qquad w \in W, \qquad (19.11)$$

where $\hat{g}(w)$ is the right hand side of (19.10) and $\hat{r}(w)$ is an arbitrary rational function which is holomorphic on $\partial W = \Gamma_\pi \cup \Gamma_{\pi-2\phi}$. For example, let us set $\hat{r}(w) \equiv 0$ for brevity. Similarly, we will drop the sum in (19.10), so finally, we set

$$\hat{T}(w)\hat{v}_{21}(w) = -\frac{1}{2\pi i} \langle p.f. \frac{\hat{T}^-(w')\hat{G}^-(w')}{\hat{Q}_1^-(w')}, \frac{e^{-(w-w')^2}}{w-w'} \rangle_{\Gamma_\pi}, \qquad w \in W. \qquad (19.12)$$

V. Further we can follow all the steps of the previous section and obtain the functions $v_{l\beta}(z_l)$ which are holomorphic in V_l^+ under conditions (19.2)–(19.4), and invariant with respect to the covering automorphisms h_l. It remains to prove (19.9) under a finite number of orthogonality conditions. First, let us consider the function

$$\psi(w) := p.f. \frac{\hat{T}^-(w)\hat{G}^-(w)}{\hat{Q}_1^-(w)}, \qquad w \in \Gamma_\pi. \qquad (19.13)$$

In our case $f_1^0 \in H^{s-m_1-1/2}(\mathbb{R})$, and hence,

$$\int_{\Gamma_0 \cup \Gamma_\pi} e^{(2s-2m_1)|w|} |\hat{f}_1^0(w)|^2 dw < \infty. \qquad (19.14)$$

Let us drop all constants in (9.9) for the simplicity of calculations. Then

$$F(z) \equiv -A_1^0(z)\tilde{B}_{20}(z_2)\tilde{f}_1^0(z_1) \qquad (19.15)$$

by (12.8), since $f_2^0 = 0$. Now (14.9) gives

$$G := S_1^{h_1} F - S_1 F^{h_1} = -[S_1(h_1 z)A_1^0(z)\tilde{B}_{20}(z) - S_1(z)A_1^0(h_1 z)\tilde{B}_{20}(h_1 z)]\tilde{f}_1^0(z_1). \qquad (19.16)$$

Lemma 19.3 *For large $R > 0$ we have*

$$\psi(w) = M(w)\hat{f}_1^0(w - i0) \text{ for } w \in \Gamma_\pi \text{ with } |w| > R, \quad \text{where} \quad |M(w)| \le Ce^{(1+\tau-m_1)|w|} \qquad (19.17)$$

and $\tau = \operatorname{Re}\mu/\nu$.

Proof Equations (14.9), (19.6) and (12.9) imply that, for large $R > 0$,

$$|\hat{Q}_1(w)| \ge \varkappa e^{2(m_1+m_2)|w|}, \qquad |\operatorname{Re} w| > R, \quad w \in W, \qquad (19.18)$$

where $\varkappa > 0$. Moreover, (17.21) and (16.4) imply that

$$|\hat{T}(w)| \sim \varkappa_\pm e^{\tau|w|}, \qquad \pm\operatorname{Re} w \to \infty, \quad w \in W, \qquad (19.19)$$

where $\varkappa_\pm > 0$. Finally, substituting (19.16) into (19.13) and using (19.18), (19.19) and (10.10), we obtain (19.17) for large $R > 0$. $\square$

Now (19.14) implies that

$$\int_{w \in \Gamma_\pi : |w| > R} e^{(2s-2-2\tau)|w|} |\psi(w)|^2 dw < \infty. \qquad (19.20)$$

VI. Let us denote the Cauchy type integral

$$I(w) := \langle \psi(w'), \frac{e^{-(w-w')^2}}{w - w'} \rangle_{\Gamma_\pi}, \qquad w \in W. \qquad (19.21)$$

Lemma 19.4 *For $a \in (0, \pi)$ we have*

$$\int_{\Gamma_a} e^{(2s-2-2\tau)|w|} |I(w)|^2 dw < \infty. \qquad (19.22)$$

Proof Let us split $\psi(w) = \zeta(w)\psi(w) + (1 - \zeta(w))\psi(w)$ for $w \in \Gamma_\pi$, where $\zeta \in C_0^\infty(\Gamma_\pi)$ and

$$\zeta(w) = \left. \begin{cases} 1 & \text{for } w \in \Gamma_\pi, \; |w| \le R \\ 0 & \text{for } w \in \Gamma_\pi, \; |w| \ge R + 1 \end{cases} \right|. \tag{19.23}$$

Then

$$\left| \langle \zeta(w')\psi(w'), \frac{e^{-(w-w')^2}}{w - w'} \rangle_{\Gamma_\pi} \right| \le C(a)e^{-\varepsilon|w|^2}, \qquad w \in \Gamma_a, \tag{19.24}$$

where $\varepsilon > 0$. It remains to show estimate (19.22) for the function

$$I_1(w) = \int_{\Gamma_\pi} (1 - \zeta(w'))\psi(w') \frac{e^{-(w-w')^2}}{w - w'} dw', \qquad w \in W. \tag{19.25}$$

Let us denote by $\mathcal{E}^p$ the weighted space of measurable functions $f(\xi)$ on $\mathbb{R}$ with finite norm

$$\|f\|_p^2 := \int_{\mathbb{R}} e^{2p|\xi|} |f(\xi)|^2 d\xi < \infty. \tag{19.26}$$

Writing $w = \xi + ia$ and $w' = \eta + i\pi$, estimate (19.22) reduces to the boundedness of the integral operator

$$f \mapsto Kf(\xi) = \int_{\mathbb{R}} K(\xi - \eta)f(\eta)d\eta, \qquad \text{where} \quad K(\sigma) := \frac{e^{-(\sigma + i(a-\pi))^2}}{\sigma + i(a - \pi)}. \tag{19.27}$$

This operator is bounded in $\mathcal{E}^p$ for any $a \in \mathbb{R} \setminus 0$ and $p \in \mathbb{R}$. For $p = 0$ this follows by the Fourier transform. For $p > 0$ the proof reduces to the case $p = 0$ by the estimate

$$e^{p|\xi|} |Kf(\xi)| \le \int_{\mathbb{R}} e^{p|\xi - \eta|} |K(\xi - \eta)| e^{p|\eta|} |f(\eta)| d\eta. \tag{19.28}$$

For $p < 0$ the boundedness follows by duality. Now (19.22) for the function (19.25) follows from (19.20). □

Now (19.12), (19.22) and (19.19) imply that

$$\int_{\Gamma_a} e^{(2s-2)|w|} |\hat{v}_{21}(w)|^2 dw < \infty \tag{19.29}$$

for $a \in (0, \pi)$. Hence, taking $a = \pi - \Phi$ we obtain the bound for the integral (19.9) over $\Gamma_{\pi-\Phi}$ with $\beta = 1$.

VII. We should still estimate the integral over $\Gamma_{-\Phi}$ in (19.9).

Lemma 19.5 *Bound (19.29) holds for each $a \in (-2\Phi, \pi)$ if $a \neq \pi - 2\Phi, \pi - 4\Phi, \ldots$ and the line Γ_a does not contain zeros of $\hat{Q}_2(w + 2i\Phi)$.* $\square$

Proof For $a \in (\pi - 4\Phi, \pi - 2\Phi)$ we use the representation (16.3):

$$\hat{v}_{21}(w) = \frac{\hat{Q}_1(w + 2i\Phi)}{\hat{Q}_2(w + 2i\Phi)} \hat{v}_{21}(w + 2i\Phi) - \frac{\hat{G}(w + 2i\Phi)}{\hat{Q}_2(w + 2i\Phi)}, \qquad \text{Im } w \in (\pi - 4\Phi, \pi - 2\Phi).$$

$$(19.30)$$

Bound (19.29) with $a \in (\pi - 4\Phi, \pi - 2\Phi)$ for the first term on the right-hand side follows from (19.29) with $a \in (0, \pi)$, since the quotient $\hat{Q}_1/\hat{Q}_2$ is a bounded function on Γ_a by the estimate of type (19.18) for $\hat{Q}_2$. To prove (19.29) for the second term, we first note that (19.16) gives

$$\frac{\hat{G}(w + 2i\Phi)}{\hat{Q}_2(w + 2i\Phi)} = M(w)\hat{f}_1^0(w + 2i\Phi) \text{ for } w \in \Gamma_a, \quad \text{where} \quad |M(w)| \leq C(a)e^{(1-m_1)|w|}$$

$$(19.31)$$

similarly to (19.17). Second, the bound (19.14) implies that, for any $b \in (0, \pi)$,

$$\int_{\Gamma_b} e^{(2s-2m_1)|w|} |\hat{f}_1^0(w)|^2 dw < \infty.$$

$$(19.32)$$

This follows from the analyticity of $\hat{f}_1^0(w)$ in the strip Π_0^π by the arguments from the proof of Lemma 19.4. For $a < \pi - 4\Phi$ the proof follows by induction. $\square$

Corollary 19.6 *For large $R > 0$*

$$\int_{w \in \Gamma_a : |\text{Re } w| > R} e^{(2s-2)|w|} |\hat{v}_{21}(w)|^2 dw < \infty, \qquad a \in (-2\Phi, \pi).$$

$$(19.33)$$

Indeed, Lemma 19.5 implies this bound for all $a \in (-2\Phi, \pi)$ except for a discrete set. For the remaining values of a the bound follows by arguments from the proof of the same lemma.

Estimate (19.33) with $a = -\Phi$ implies the estimate (19.9) with $\beta = 1$ under a finite number of orthogonality conditions of type (19.2)–(19.4), which remove the poles of $\hat{v}_{21}$ on $\Gamma_{-\Phi}$. Then (19.9) also holds for the symmetrization (15.13).

VIII. Next step is to prove (19.8) with $\beta = 1$. The function v_{11} can be expressed from (14.5):

$$\hat{v}_{11}(z) = \frac{\hat{F}(z) - \hat{S}_2(z)\hat{v}_{21}(z)}{\hat{S}_1(z)}, \qquad w \in \hat{V}^+.$$

$$(19.34)$$

This function is meromorphic in $\hat{V}^+$ and analytic in $\hat{V}_1^+$ under conditions (19.3). Moreover, it is invariant with respect to $\hat{h}_1$ by Lemma 14.4. Hence, it suffices to bound the integral (19.8) with $\beta = 1$ only over Γ_0 since $\Gamma_\pi = \hat{h}_1 \Gamma_0$.

This integral over large $|\text{Re } w| > R$ is finite by (19.33) and (19.14) since

$$\frac{\hat{F}(z)}{S_1(z)} = M(w)\hat{f}_1^0(w) \text{ for } w \in \Gamma_0, \quad \text{where} \quad |M(w)| \le Ce^{(1-m_1)|w|} \quad (19.35)$$

similarly to (19.31). Finally, the integral over the bounded interval $|\text{Re } w| < R$ is finite under additional orthogonality conditions removing the poles of $\hat{v}_{11}(z)$ on Γ_0 with $\text{Re } w \in [-R, R]$.

IX. The bounds (19.8) and (19.9) with $\beta = 0$ follow from the same bounds with $\beta = 1$ by (12.5) and (19.6).

As a conclusion, we obtain the following theorem.

Theorem 19.7 *Let $A(\partial)$ be the strongly elliptic Helmholtz operator (7.4) with $\omega \in i\mathbb{R}^+$ and let the Shapiro-Lopatinski condition (7.22) hold, $s > \max(3/2, m_1 + 1/2, m_2 + 1/2)$ and $s \notin \mathbb{N}+1/2$. Then in the case (12.4) solutions $u \in H^s(K)$ to the problem (7.1), (7.2) exist for a linear space of boundary data $f_l \in H^{s-m_l-1/2}(\mathbb{R}^+)$, $l = 1, 2$ satisfying a finite number of orthogonality conditions*

$$L_j(f_1, f_2) = 0, \qquad j = 1, \dots, J, \quad (19.36)$$

where L_j are linear functionals on $H^{s-m_1-1/2}(\mathbb{R}^+) \oplus H^{s-m_2-1/2}(\mathbb{R}^+)$. The space of solutions $u \in H^s(K)$ to the problem (7.1), (7.2) with $f_1 = f_2 = 0$ is finite-dimensional. $\square$

Remark 19.8 The construction above gives the Cauchy data $v_l^\beta \in \mathring{H}^{s-\beta-1/2}(\mathbb{R}^+)$ under a finite number of orthogonality conditions. The corresponding solution u is given by the formulas (8.5) and (8.3), (8.11). This solution belongs to the Sobolev space $H^s(K)$.

Moreover, our method gives also more general Cauchy data $\mathbf{v}_l^\beta \in H^{s-\beta-1/2}(\mathbb{R}^+)$ which differ from v_l^β by another choice of "arbitrary constants" in the boundary conditions (9.9), in (16.7) and in (18.4). In this case the corresponding solution $\mathbf{u}$ generally does not belong to $H^s(K)$. For example, this is possible when the boundary conditions do not satisfy some well known compatibility conditions at the edge of the wedge. On the other hand, the difference $\mathbf{u} - u$ belongs to a finite-dimensional subspace of functions given by explicit formulas. These formulas allow us to select all solutions from the Sobolev space $H^s(K)$, thus taking into account the compatibility conditions.

For $s > 3/2$ the corresponding compatibility conditions [52, (1.10)] were found for general boundary value problems in angles.

Remark 19.9 We have proved all results of Chaps. 15–19 for the Helmholtz operator (7.4) with $\omega \in i\mathbb{R}^+$. The extension to general strongly elliptic operator (7.6) is straightforward, see [52] for details.

Namely, in this more general setting our rectilinear strips $\hat{V}_l^+$, Π_a^b, etc., become curvilinear strips as in Fig. 15.2, and the interval I in the Riemann-Hilbert problem (16.6) becomes a curve connecting the points ± 1, see [52, Lemma 3.2]. In particular, Theorems 3.1 and 2.1 of [52] extend our Lemmas 15.2 and 15.4.

The reduction to the Riemann-Hilbert problem and its analysis in Chaps 16–18 remain essentially unchanged, as well as the reconstruction of solution in Chap. 19.

Remark 19.10 Theorem 7 concerns the Fredholmness of boundary value problem (7.1), (7.2). However, in many practical applications with higher order boundary operators a criterion of uniqueness is necessary, see for example [6, 13, 107, 137, 138]. The main advantage of our method is that it gives all solutions in the space of tempered distributions. This allows us to formulate the existence and uniqueness as invertibility of a finite-dimensional matrix which should be calculated in each case of interest.

Up to now this matrix is calculated explicitly only for some specific cases, see, for example, [65, 89]. In other problems the matrix can be calculated numerically.□

Chapter 20
Extension of the Method to Non-convex Angle

In this chapter we extend the approach of Chaps. 7–19 to problems (7.1)–(7.2) in angles Q of magnitude $\Phi > \pi$ using our method [58, 61]. In suitable system of coordinates the angle Q coincides with $\mathbb{R}^2 \setminus K$ where K is the first quadrant $\mathbb{R}^+ \times \mathbb{R}^+$, since $\Phi > \pi$. Then the stationary diffraction problem can be written as the b.v.p.

$$A(\partial)u(x) = \sum_{|\alpha| \leq 2} a^\alpha \partial^\alpha u(x) = 0, \qquad x \in Q, \qquad (20.1)$$

$$\begin{cases} B_1(\partial)u(x_1, 0-) = \sum_{|\alpha| \leq m_1} b_l^\alpha \partial^\alpha u(x_1, 0-) = f_1(x_1), & x_1 > 0 \\ B_2(\partial)u(0-, x_2) = \sum_{|\alpha| \leq m_2} b_l^\alpha \partial^\alpha u(0-, x_2) = f_2(x_2), & x_2 > 0 \end{cases} \qquad (20.2)$$

The boundary data $f_l(x_l) \in S'(R^+)$ are tempered distributions on $\mathbb{R}^+$. We keep the notations (7.5) and the assumption of strong ellipticity (7.6) of the operator $A(\partial)$. Now the Shapiro-Lopatinski condition (7.8) reads

$$\tilde{B}_l^0(z_1, z_2) \neq 0 \quad \text{for} \quad z \in C_l^+, \quad l = 1, 2. \qquad (20.3)$$

Here

$$C_1^+ := \{z \in \mathbb{R} \times \mathbb{C}^+ : \tilde{A}^0(z) = 0\}, \qquad C_2^+ := \{z \in \mathbb{C}^+ \times \mathbb{R} : \tilde{A}^0(z) = 0\},$$

where $\mathbb{C}^+ := \{z \in \mathbb{C} : \operatorname{Im} z > 0\}$. In the notations (7.10)–(7.18), the Shapiro-Lopatinski condition (20.3) with $l = 1$ is equivalent to the lower bound of type (7.8),

$$|\tilde{B}_1(z_1, \lambda_2^+(z_1))| \geq \varkappa_1 |z_1|^{m_1}, \qquad |z_1| > R, \quad z_1 \in \mathbb{R}$$

for sufficiently large $R > 0$ where $\varkappa_1 > 0$; a similar estimate holds for $l = 2$.

© Springer Nature Switzerland AG 2019
A. Komech, A. Merzon, *Stationary Diffraction by Wedges*, Lecture Notes in Mathematics 2249, https://doi.org/10.1007/978-3-030-26699-8_20

We will extend the methods of Chaps. 6–16 to solve the boundary problem (20.1)–(20.2) in quadratures. We will develop the method for distributional solutions as well as for regular solutions defined by (7.22) and (7.23), respectively, with Q instead of K. In particular, the definition (8.1) now becomes

$$u_0(x) = \begin{cases} u(x), & x \in Q \\ 0, & x \in \mathbb{R}^2 \setminus Q \end{cases} \Bigg| .$$

All constructions and formulas of Chaps. 7 and 8 can be extended to the problem (20.1)–(20.2) with obvious modifications: now the Cauchy data u_l^β are defined by the limits (8.6) with $0-$ instead of $0+$, and $\tilde{v}_l^\beta$ are the Fourier transforms of distributions (8.10). Further, the sign of the right-hand sides of the formulas (8.9) and (8.11) should be changed, and formulas (8.2)–(8.5) hold with the distribution γ given by negative right hand side of (10.8).

For definiteness, we will consider below the strongly elliptic Helmholtz operator (7.4) with $\mathrm{Im}\,\omega > 0$. The case $\mathrm{Im}\,\omega < 0$ can be considered similarly. The extension to general strongly elliptic operators A is straightforward.

Similarly to (10.7)–(10.10) we have

$$\tilde{\gamma}(z) = \tilde{\gamma}_1(z) + \tilde{\gamma}_2(z), \qquad z \in \mathbb{R}^2,$$

where

$$\begin{cases} \tilde{\gamma}_1(z) = \tilde{v}_1^0(z_1)(iz_2 - 2iz_1 \cos \Phi) - \tilde{v}_1^1(z_1) \\ \tilde{\gamma}_2(z) = \tilde{v}_2^0(z_2)(iz_1 - 2iz_2 \cos \Phi) - \tilde{v}_2^1(z_1) \end{cases} \Bigg| , \qquad z \in \mathbb{R}^2. \qquad (20.4)$$

The Riemann manifold V is defined by (10.3) as in the case $\Phi < \pi$, and the functions $\tilde{v}_l^\beta(z_l)$ and $\tilde{\gamma}_l(z)$ are analytic in the same domains V_l^+ defined by (12.1). The last two equations (12.3) hold as in the case $\Phi < \pi$.

All three lemmas of Chap. 1 remain valid with their proofs. All arguments of Chaps. 9 and 11–19 also admit obvious modifications. The only difference is the "Connection Equation" (10.7) on the Riemann surface obtained in Chap. 9 for the case $\Phi < \pi$.

20.1 Connection Equation for Non-convex Angle

We will show that the Connection Equation (10.7) in the case $\Phi > \pi$ hold *for analytic continuation* of the functions $\tilde{v}_l^\beta$ *along the Riemann surface V*. The inclusion supp $u_0 \subset Q$ means, by definition, that

$$\langle u_0, \psi \rangle = 0 \qquad \text{for} \quad \psi \in \mathcal{S}(K) := \{\psi \in \mathcal{S}(\mathbb{R}^2) : \text{supp } \psi \subset \overline{K}\}, \qquad (20.5)$$

where $K := \mathbb{R}^+ \times \mathbb{R}^+$ is the first quadrant, and the brackets $\langle \cdot, \cdot \rangle$ denote various extensions of the bilinear quadratic form $\int_{\mathbb{R}^2} f(x)g(x)dx$ with $f, g \in L^2(\mathbb{R}^2)$. Our plan is as follows.

1. At first, we will transform the relation (20.5) into an *integral identity* for the functions $\tilde{\gamma}_l$ on the Riemann surface using the Cauchy residue theorem.
2. Second, we will derive from this integral identity an analytic continuation of the functions $\tilde{\gamma}_l$ which satisfy the "Connection Equation" (20.7) similar to (10.5).

We keep the notation of (12.1) and set

$$V_* := V_1^- \cap V_2^-, \tag{20.6}$$

see Fig. 10.1. This intersection is nonempty. Indeed, (7.17) implies that this intersection contains, in particular, the points $(z_1 - i\varepsilon, \lambda_2^-(z_1 - i\varepsilon))$ with $z_1 \in \mathbb{R}$ and small $\varepsilon > 0$. The main result of this section is the following theorem.

Theorem 20.1 *Let $A(\partial)$ be the operator (7.4) with $\operatorname{Im}\omega > 0$, and $u \in S'(Q)$ be any distributional solution to (20.1). Then the corresponding functions (20.4) admit an analytic continuation along the Riemann surface V from V_1^+ into V_*, and*

$$\tilde{\gamma}_1(z) + \tilde{\gamma}_2(z) = 0, \qquad z \in V_* \tag{20.7}$$

for these analytic continuations. $\square$

20.2   Integral Connection Equation on the Riemann Surface

After the Fourier transform, identity (20.5) reads

$$\langle \tilde{u}_0(z), \hat{\psi}(-z) \rangle = \langle \frac{\tilde{\gamma}(z)}{\tilde{A}(z)}, \hat{\psi}(-z) \rangle = \langle \frac{\tilde{\gamma}_1(z)}{\tilde{A}(z)}, \hat{\psi}(-z) \rangle + \langle \frac{\tilde{\gamma}_2(z)}{\tilde{A}(z)}, \hat{\psi}(-z) \rangle = 0, \qquad \psi \in S(K).$$

$$\tag{20.8}$$

By (20.4) the functions $\tilde{\gamma}_1$ and $\tilde{\gamma}_2$ are analytic in $z \in \mathbb{C}^+ \times \mathbb{C}$ and $\mathbb{C} \times \mathbb{C}^+$, respectively, and

$$\begin{cases} |\tilde{\gamma}_1(z)| \le C \frac{(|z|+1)^M}{|\operatorname{Im} z_1|^N}, & z \in \mathbb{C}^+ \times \mathbb{C} \\[2mm] |\tilde{\gamma}_2(z)| \le C \frac{(|z|+1)^M}{|\operatorname{Im} z_2|^N}, & z \in \mathbb{C} \times \mathbb{C}^+ \end{cases} \tag{20.9}$$

by the Paley-Wiener theorem. Let us choose an appropriate class of exponentially decaying test functions.

Definition 20.2

(i) S_δ with $\delta > 0$ denotes the space of test functions $\psi \in S(K)$ with finite seminorms

$$\sup_{x \in K} |e^{\delta(x_1+x_2)} x^\beta \partial^\alpha \psi(x)| < \infty, \qquad \forall \, \alpha, \beta.$$

(ii) $\tilde{S}_\delta$ denotes the space of Fourier transforms $\tilde{\psi}$ of functions $\psi \in S_\delta$. $\qquad \square$

We will need a characterization of the Fourier transforms of these test functions. Let us denote

$$\mathbb{C}(-\delta) := \{z \in \mathbb{C} : \mathrm{Im}\, z > -\delta\}.$$

Lemma 20.3 *The inclusion $\psi \in S_\delta$ is equivalent to the bounds*

$$\sup_{z \in \mathbb{C}(-\delta) \times \mathbb{C}(-\delta)} |\partial^\beta z^\alpha \tilde{\psi}(z)| < \infty, \qquad \forall \, \alpha, \beta. \tag{20.10}$$

Proof This lemma follows from the Schwartz–Paley–Wiener theorem [115, 122]. $\square$

Strong ellipticity (7.6) implies that there exists $\varepsilon \in (0, \delta)$ such that

$$\tilde{A}(z) \neq 0 \qquad \text{for} \quad |\mathrm{Im}\, z| \leq \varepsilon. \tag{20.11}$$

Hence, (20.8) can be written as

$$\int_{\Gamma^\varepsilon \times \mathbb{R}} \frac{\tilde{\gamma}_1(z)}{\tilde{A}(z)} \tilde{\psi}(-z) dz + \int_{\mathbb{R} \times \Gamma^\varepsilon} \frac{\tilde{\gamma}_2(z)}{\tilde{A}(z)} \tilde{\psi}(-z) dz = 0, \tag{20.12}$$

where

$$\Gamma^\varepsilon := \{\zeta \in \mathbb{C} : \mathrm{Im}\, \zeta = \varepsilon\}. \tag{20.13}$$

Let us calculate these integrals using the Cauchy residue theorem. The symbol of the Helmholtz operator (7.4) factorizes as follows:

$$\tilde{A}(z) = -z_1^2 - z_2^2 + 2z_1 z_2 \cos \Phi - \omega^2 \sin^2 \Phi = -(z_1 - \lambda_1^+(z_2))(z_1 - \lambda_1^-(z_2))$$

$$= -(z_2 - \lambda_2^+(z_1))(z_2 - \lambda_2^-(z_1)), \quad |\mathrm{Im}\, z| \leq \varepsilon.$$

Inequalities (20.11) and (7.17) imply that

$$\pm \operatorname{Im} \lambda_1^{\pm}(z_2) > 0 \quad \text{for} \quad |\operatorname{Im} z_2| \le \varepsilon, \qquad \text{and} \qquad \pm \operatorname{Im} \lambda_2^{\pm}(z_1) > 0 \quad \text{for} \quad |\operatorname{Im} z_1| \le \varepsilon. \tag{20.14}$$

The roots are given by

$$\lambda_1^{\pm}(z_2) = z_2 \cos \Phi \mp \sin \Phi \sqrt{\omega^2 - z_2^2}, \qquad \lambda_2^{\pm}(z_1) = z_1 \cos \Phi \mp \sin \Phi \sqrt{\omega^2 - z_1^2},$$

since $\sin \Phi < 0$ and we choose the analytic branch of the square root with

$$\operatorname{Im} \sqrt{\omega^2 - \zeta^2} > 0 \qquad \text{for} \quad |\operatorname{Im} \zeta| < \varepsilon. \tag{20.15}$$

Such branch exists for $0 < \varepsilon < \operatorname{Im} \omega$ (we consider the case $\operatorname{Im} \omega > 0$). Hence, we can calculate the first integral (20.12) closing the contour of integration $z_2 \in \mathbb{R}$ into lower complex half-plane and using the estimates (20.9) and the rapid decay (20.10):

$$I_1 = - \int_{\Gamma^{\varepsilon} \times \mathbb{R}} \frac{\tilde{\gamma}_1(z)}{(z_2 - \lambda_2^{-}(z_1))(z_2 - \lambda_2^{+}(z_1))} \tilde{\psi}(-z) dz$$

$$= 2\pi i \int_{\Gamma^{\varepsilon}} \frac{\tilde{\gamma}_1(z_1, \lambda_2^{-}(z_1))}{\lambda_2^{-}(z_1) - \lambda_2^{+}(z_1)} \tilde{\psi}(-z_1, -\lambda_2^{-}(z_1)) dz_1$$

$$= -\frac{\pi i}{\sin \Phi} \int_{\Gamma^{\varepsilon}} \frac{\tilde{\gamma}_1(z_1, \lambda_2^{-}(z_1))}{\sqrt{\omega^2 - z_1^2}} \tilde{\psi}(-z_1, -\lambda_2^{-}(z_1)) dz_1$$

Similarly, the second integral (20.12) for $\varepsilon < |\operatorname{Im} \omega|$ equals

$$I_2 = -\frac{\pi i}{\sin \Phi} \int_{\Gamma^{\varepsilon}} \frac{\tilde{\gamma}_2(\lambda_1^{-}(z_2), z_2)}{\sqrt{\omega^2 - z_2^2}} \tilde{\psi}(-\lambda_1^{-}(z_2), -z_2) dz_2$$

Now (20.12) reads

$$\int_{\Gamma^{\varepsilon}} \frac{\tilde{\gamma}_1(z_1, \lambda_2^{-}(z_1))}{\sqrt{\omega^2 - z_1^2}} \tilde{\psi}(-z_1, -\lambda_2^{-}(z_1)) dz_1 + \int_{\Gamma^{\varepsilon}} \frac{\tilde{\gamma}_2(\lambda_1^{-}(z_2), z_2)}{\sqrt{\omega^2 - z_2^2}} \tilde{\psi}(-\lambda_1^{-}(z_2), -z_2) dz_2 = 0.$$

Equivalently,

$$\int_{\Gamma_1^{\varepsilon}} \frac{\tilde{\gamma}_1(z)}{\sqrt{\omega^2 - z_1^2}} \tilde{\psi}_{-}(z) dz_1 + \int_{\Gamma_2^{\varepsilon}} \frac{\tilde{\gamma}_2(z)}{\sqrt{\omega^2 - z_2^2}} \tilde{\psi}_{-}(z) dz_2 = 0, \tag{20.16}$$

where $\tilde{\psi}_-(z) := \tilde{\psi}(-z)$, and we denote the contours on the Riemann surface V,

$$\Gamma_1^\varepsilon := \{z = (z_1, \lambda_2^-(z_1)) : z_1 \in \Gamma^\varepsilon)\} \subset V_2^-, \quad \Gamma_2^\varepsilon := \{z = (\lambda_1^-(z_2), z_2) : z_2 \in \Gamma^\varepsilon)\} \subset V_1^-.$$
(20.17)

20.3 Lifting onto the Universal Covering

Let us rewrite the identity (20.16) in the coordinate w on the universal covering. We should identify the covering for the regions V_1^+, V_2^+ and V_*. Now the identifications (15.2) are not appropriate, since $\hat{V}_1^+$ and $\hat{V}_2^+$ should be adjacent to $\hat{V}_* := \hat{V}_1^- \cap \hat{V}_2^-$. Hence, we choose

$$\begin{cases} \hat{V}_1^+ = \{w \in \mathbb{C} : \quad -\pi/2 < \operatorname{Im} w < 3\pi/2, \quad \operatorname{Im} z_1(w) > 0\}, \\ \hat{V}_2^+ = \{w \in \mathbb{C} : -\pi/2 - \Phi < \operatorname{Im} w < 3\pi/2 - \Phi, \operatorname{Im} z_2(w) > 0\} \end{cases}.$$

Similarly,

$$\begin{cases} \hat{V}_1^- = \{w \in \mathbb{C} : \quad -3\pi/2 < \operatorname{Im} w < \pi/2, \quad \operatorname{Im} z_1(w) < 0\} \\ \hat{V}_2^- = \{w \in \mathbb{C} : \pi/2 - \Phi < \operatorname{Im} w < 5\pi/2 - \Phi, \operatorname{Im} z_2(w) < 0\} \end{cases}.$$

Respectively, we denote

$$\hat{V}_* := \hat{V}_1^- \cap \hat{V}_2^-$$

according to (20.6), see Fig. 20.1. Changing the variables according to (15.1), we obtain by (20.15),

$$\begin{cases} dz_1 = -i\omega \cosh w\, dw = -i\sqrt{\omega^2 - z_1^2}\, dw, \quad w \in \hat{\Gamma}_1^\varepsilon \\[2mm] dz_2 = -i\omega \cosh(w + i\Phi)\, dw = i\sqrt{\omega^2 - z_2^2}\, dw, \ w \in \hat{\Gamma}_2^\varepsilon \end{cases},$$

where the branches of the square roots are chosen according to (20.15). Hence, (20.16) becomes

$$\int_{\hat{\Gamma}_1^\varepsilon} \hat{\tilde{\gamma}}_1(w)\hat{\tilde{\psi}}_-(w)dw - \int_{\hat{\Gamma}_2^\varepsilon} \hat{\tilde{\gamma}}_2(w)\hat{\tilde{\psi}}_-(w)dw = 0.$$
(20.18)

Let us denote by V_*^δ the neighborhood of $V_* := V_1^- \cap V_2^-$ in V bounded by the contours Γ_1^δ and Γ_2^δ which are defined similarly to (20.17), see Fig. 20.2. The function $\tilde{\psi}(-z)$ is analytic for $\operatorname{Im} z \in (-\infty, \delta) \times (-\infty, \delta)$ by Lemma 20.3. Hence, the restriction of $\tilde{\psi}(-z)$ onto the Riemann manifold V is analytic in V_*^δ.

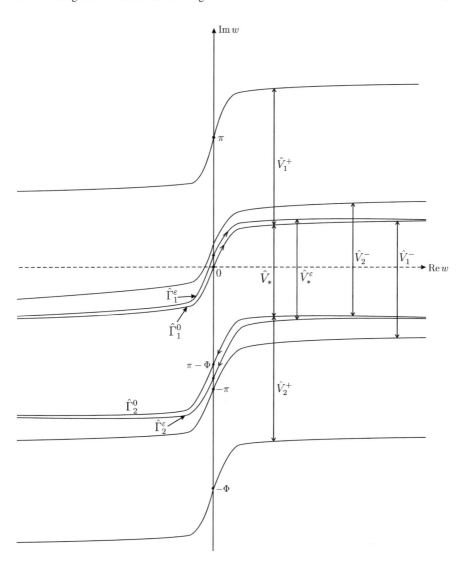

Fig. 20.1 Contours of integration

Respectively, the lifting $\hat{\tilde{\psi}}_{-}(w)$ is analytic in the region $\hat{V}_{*}^{\delta}$ bounded by the contours $\hat{\Gamma}_{1}^{\delta}$ and $\hat{\Gamma}_{2}^{\delta}$ and containing $\hat{V}_{*}$, see Fig. 20.2. Hence, the identity (20.18) suggests the following lemma.

Lemma 20.4 *Traces* $\hat{\tilde{\gamma}}_{1}(w)|_{\hat{\Gamma}_{1}^{\varepsilon}}$ *and* $-\hat{\tilde{\gamma}}_{2}(w)|_{\hat{\Gamma}_{2}^{\varepsilon}}$ *are boundary values of an analytic function in the region* $\hat{V}_{*}^{\varepsilon} \subset \hat{V}_{*}^{\delta}$ *bounded by the contours* $\hat{\Gamma}_{1}^{\varepsilon}$ *and* $\hat{\Gamma}_{2}^{\varepsilon}$.   □

This lemma obviously implies Theorem 20.1.

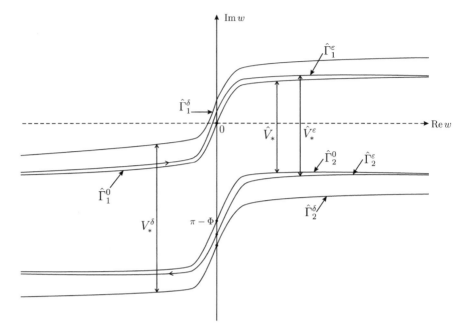

Fig. 20.2 The analytic continuation

20.4   The Cauchy Kernel on the Riemann Surface

To prove Lemma 20.4 we will construct a special Cauchy kernel on the Riemann surface.

Proposition 20.5 *Let $\delta \in (0, |\operatorname{Im} \omega|)$ and $\varepsilon \in (0, \delta)$. Then*

(i) *There exists an analytic function $K(w', w)$ of $(w', w) \in \hat{V}_*^{\delta} \times \hat{\Gamma}$ with $w' \neq w$, where $\hat{\Gamma} := \hat{\Gamma}_1^{\varepsilon} \cup \hat{\Gamma}_2^{\varepsilon}$, such that*

$$|K(w', w)| \leq \frac{C(w') \exp(e^{-|w|/5})}{|w' - w|}, \qquad (w', w) \in \hat{V}_*^{\delta} \times \hat{\Gamma}.$$

(ii) *The following identity holds*

$$\int_{\hat{\Gamma}_1^{\varepsilon}} \hat{\bar{\gamma}}_1(w) K(w', w) dw - \int_{\hat{\Gamma}_2^{\varepsilon}} \hat{\bar{\gamma}}_2(w) K(w', w) dw = 0, \qquad w' \in \hat{V}_*^{\delta} \setminus \hat{V}_*^{\varepsilon}.$$
$$(20.19)$$

(iii) *For w' in a neighborhood of $\hat{\Gamma}$ we have*

$$K(w', w) = \frac{1}{2\pi i (w - w')} + H(w', w), \qquad (20.20)$$

where $H(w', w)$ is an analytic function for $w' = w$.   □

This proposition immediately implies Lemma 20.4 and Theorem 20.1. Namely, let us denote by $f(w')$ the left hand side of (20.19) for $w' \in \hat{V}_*^\delta \setminus \hat{\Gamma}$. Then Plemelj's formulas imply that the boundary values of $f|_{\hat{V}_*^\varepsilon}$ read

$$
f(w') = \begin{cases} -\hat{\tilde{\gamma}}_1(w'), & w' \in \hat{\Gamma}_1^\varepsilon \\[2mm] +\hat{\tilde{\gamma}}_2(w'), & w' \in \hat{\Gamma}_2^\varepsilon \end{cases} \tag{20.21}
$$

by (20.19). Hence, Lemma 20.4 follows: f is the analytic continuation of $-\hat{\tilde{\gamma}}_1$ and of $\hat{\tilde{\gamma}}_2$ into $\hat{V}_\varepsilon$. In other words, $\hat{\tilde{\gamma}}_1$ and $\hat{\tilde{\gamma}}_2$ admit an analytic continuation in $\hat{V}_*^\varepsilon$, and

$$
\hat{\tilde{\gamma}}_1(w) + \hat{\tilde{\gamma}}_2(w) = 0, \qquad \hat{V}_\sigma := \hat{V}_1^+ \cup \hat{V}_2^+ \cup \overline{\hat{V}_*},
$$

which implies (20.7).

It remains to prove Proposition 20.5. First, let us construct appropriate test functions $\psi \in S_\varepsilon$ with $\varepsilon \in (0, 1)$. Let us define

$$
\Lambda(z) := e^{-(z_1+i)^{1/5}(z_2+i)^{1/5}}, \qquad z \in \mathbb{C}(-\varepsilon) \times \mathbb{C}(-\varepsilon),
$$

where we choose the branch with $\arg \zeta^{1/5} \in (0, \pi/5)$ for $\mathrm{Im}\, \zeta > 0$. The simplest example of $\psi \in S_\varepsilon$ is defined by its Fourier transform

$$
\tilde{\psi}(z) := \frac{\Lambda(z)}{(z_1 + z_1')(z_2 + z_2')}, \qquad z \in \mathbb{C}(-\varepsilon) \times \mathbb{C}(-\varepsilon),
$$

where $\mathrm{Im}\, z_1'$, $\mathrm{Im}\, z_2' > \varepsilon$. Then estimates (20.10) hold, and hence, $\psi \in S_\varepsilon$ by Lemma 20.3. The corresponding function $\tilde{\psi}_-$ reads

$$
\tilde{\psi}_-(z) = \frac{\Lambda(-z)}{(-z_1 + z_1')(-z_2 + z_2')} \qquad \text{for} \quad \mathrm{Im}\, z_1 < \varepsilon, \ \mathrm{Im}\, z_2 < \varepsilon, \tag{20.22}
$$

and identity (20.16) holds for this function since $\mathrm{Im}\, z_1'$, $\mathrm{Im}\, z_2' > \varepsilon$. By (15.1) the lifting of this function reads

$$
\hat{\tilde{\psi}}_-(w) = \frac{\hat{\Lambda}(w)}{(i\omega \sinh w + z_1')(i\omega \sinh(w + i\Phi) + z_2')}, \qquad w \in \hat{V}_*^\varepsilon. \tag{20.23}
$$

Let us recall that $\mathrm{Im}\, \zeta|_{\Gamma^\varepsilon} \equiv \varepsilon$ by the definition (20.13). Therefore,

$$
\mathrm{Im}\, z_1 \equiv \varepsilon \quad \text{for} \quad z = (z_1, z_2) \in \Gamma_1^\varepsilon \quad \text{and} \quad \mathrm{Im}\, z_2 \equiv \varepsilon \quad \text{for} \quad z = (z_1, z_2) \in \Gamma_2^\varepsilon
$$

by the definition (20.17). Hence,

$$
\begin{cases}
\operatorname{Im} z_1 > \varepsilon \text{ for } z = (z_1, z_2) \in [V_*^\delta \setminus V_*^\varepsilon] \cap V_1^+ \\
\operatorname{Im} z_2 > \varepsilon \text{ for } z = (z_1, z_2) \in [V_*^\delta \setminus V_*^\varepsilon] \cap V_2^+
\end{cases},
$$

see Fig. 20.2. Therefore, identity (20.16) holds for the function (20.22) if z_1' is the first coordinate of a point $z' = (z_1', z_2') \in [V_*^\delta \setminus V_*^\varepsilon] \cap V_1^+$ and z_2' is the second coordinate of a point $z' = (z_1', z_2') \in [V_*^\delta \setminus V_*^\varepsilon] \cap V_2^+$. Respectively, by (15.1) the identity (20.18) holds for the function (20.23) if $z_1' = -i\omega \sinh w'$ with $w' \in [\hat{V}_*^\delta \setminus \hat{V}_*^\varepsilon] \cap \hat{V}_1^+$ and $z_2' = -i\omega \sinh(w' + i\Phi)$ with $w' \in [\hat{V}_*^\delta \setminus \hat{V}_*^\varepsilon] \cap \hat{V}_2^+$. Hence, (20.19) holds for the kernel

$$
K(w', w) := \frac{1}{2\pi i} \frac{\hat{\Lambda}(w)}{\hat{\Lambda}(w')} \frac{\cosh w' \cosh(w' + i\Phi)}{(\sinh w - \sinh w')(\sinh(w + i\Phi) - \sinh(w' + i\Phi))},
$$

which is proportional to the function (20.23), and the factors provide the splitting (20.20).

20.5   Reduction to the Riemann–Hilbert Problem

For $\Phi < \pi$ we have obtained the Connection Equation in Chap. 10. In Chaps. 11–16 we reduced this equation to the Riemann-Hilbert problem, which was solved in Chaps. 17–19. This reduction in the case $\Phi > \pi$ can be done by the same methods with nonessential modifications.

Namely, in the case $\Phi > \pi$ we cannot express the Connection Equation (20.7) in the Cauchy data $v_{l\beta}(z)$ like the first equation of (12.3), since now the Cauchy data generally do not admit an analytic continuation into V_*, in contrast to the functions $\gamma_l(z)$. However, such analytic continuation is possible when we consider the cases $f_1 = 0$ and $f_2 = 0$.

For example, in the case $f_2 = 0$ and under condition (12.4) we can eliminate $v_{20}(z)$ by (12.5), where g_2 is a polynomial from (9.9). Hence, (20.7) implies that $v_{21}(z)$ admits an analytic continuation into V_* and even in V_1^+. Finally, eliminating $v_{10}(z)$ by (12.5), we obtain an equation of type (14.5). Now all further steps of our method from Chaps. 14–19 admit suitable extensions.

Chapter 21
Comments

1. The MAF method for solution of the boundary value problem (7.1), (7.2) in angles $\Phi < \pi$ was introduced in [52] for general strongly elliptic operators (7.6). We give its complete presentation in Chaps. 7–19 for the most important case of the Helmholtz operator (7.4) with complex $\omega \notin \mathbb{R}$.

   This presentation also contains many important details and results, which we publish for the first time. In particular, Lemma 7.1 is new, as well as all results of Chap. 8 which concern distributional solutions.

   Theorem 19.2 is new. Theorem 19.7 was formulated in [52, Theorem 6.1], however, its proof was rather sketchy. We give its complete proof in Sect. 19.2 of the present book for the first time.

2. The reduction of boundary problems in angles to the Riemann-Hilbert problem was given for the first time by Malyuzhinets in the case of the impedance (Robin) boundary condition [76, 78].

3. The extension of the MAF method to $\Phi > \pi$ was done in [58, 61].

4. The Helmholtz operator (4.18) with real ω is not strongly elliptic. In this case the geometry of the regions V_l^+ and V^* is quite different from the case $\operatorname{Im} \omega \neq 0$, and the relation of type (16.6) holds only for $t \in (0, 1)$. Therefore, this Riemann-Hilbert problem is ill-posed, as well as the problem (7.1), (7.2). This is the well-known difficulty for the Helmholtz equation in unbounded domains, where some selection rules must be added to make the problem well-posed: either the Limiting absorption principle, or the Limiting Amplitude Principle, or the Sommerfeld radiation condition.

This problem was solved by Merzon [88], who showed that the Limiting Absorption Principle makes this problem well-posed in the case of Dirichlet and Neumann boundary conditions. Namely, he proved that this principle implies the relation (16.6) with $\check{G}(t) = 0$ on the interval $(-1, 0)$.

© Springer Nature Switzerland AG 2019

A. Komech, A. Merzon, *Stationary Diffraction by Wedges*, Lecture Notes in Mathematics 2249, https://doi.org/10.1007/978-3-030-26699-8_21

Appendix A
Sobolev Spaces on the Half-Line

The following lemma implies relation (11.1), which was used in the proof of Theorem 19.7. The Sobolev space $\overset{\circ}{H}{}^s(\mathbb{R}^+) \subset H^s(\mathbb{R}^+)$ for $s > -1/2$, see [1, 134]. For $n = 0, 1, \ldots$ the differential operator $P = (\frac{d}{dx} + 1)^n : H^s(\mathbb{R}^+) \to H^{s-n}(\mathbb{R}^+)$ is continuous for all $s \in \mathbb{R}$.

Lemma A.1 *Let $s > -1/2$, $s \notin \mathbb{N} - 1/2$, and denote $n = [s + 1/2]$. Then*

$$H^s(\mathbb{R}^+) = \overset{\circ}{H}{}^s(\mathbb{R}^+) \oplus \operatorname{Ker} P, \tag{A.1}$$

and $\dim \operatorname{Ker} P = n$. $\qquad\square$

Proof To prove (A.1) we should split every function $u \in H^s(\mathbb{R}^+)$ into the sum

$$u = v + k \tag{A.2}$$

with unique $v \in \overset{\circ}{H}{}^s(\mathbb{R}^+)$ and $k \in \operatorname{Ker} P$.

(i) First, let us calculate $\operatorname{Ker} P$. Solving the differential equation $Pu = 0$ we see that u is the linear span of

$$x^k e^{-x} : 0 \le k < n.$$

Hence, $\dim \operatorname{Ker} P = n$, and $\overset{\circ}{H}{}^s(\mathbb{R}^+) \cap \operatorname{Ker} P = 0$. Thus, the splitting (A.2) is unique if it exists.

(ii) The splitting (A.2) is equivalent to the equation

$$Pv = f := Pu \in H^{s-n}(\mathbb{R}^+). \tag{A.3}$$

© Springer Nature Switzerland AG 2019
A. Komech, A. Merzon, *Stationary Diffraction by Wedges*, Lecture Notes in Mathematics 2249, https://doi.org/10.1007/978-3-030-26699-8

Here $|s - n| < 1/2$. Hence, by Stein's result [134] there exists a unique $f_0 \in \mathring{H}^{s-n}(\mathbb{R}^+)$, which is an extension of $f(x)$ by zero for $x < 0$. Let us consider the differential equation

$$(\frac{d}{dx} + 1)^n v_0(x) = f_0(x), \qquad x \in \mathbb{R}. \tag{A.4}$$

After applying the Fourier transform, $(-iz + 1)^n \tilde{v}_0(z) = \tilde{f}_0(z)$ for $z \in \mathbb{R}$, and hence,

$$\tilde{v}_0(z) \equiv i^n \frac{\tilde{f}_0(z)}{(z + i)^n}, \qquad z \in \mathbb{R}. \tag{A.5}$$

Obviously, $v_0 \in H^s(\mathbb{R})$. Finally, supp $v_0 \subset \overline{\mathbb{R}^+}$ by the Paley-Wiener theorem, since (A.5) holds also for $z \in \mathbb{C}^+$. Hence, (A.4) implies (A.3) with $v \in \mathring{H}^s(\mathbb{R}^+)$, which is the restriction of $v_0(x)$ to $x > 0$.

Now relation (11.1) follows, since $H^s(\mathbb{R}^+)/\mathring{H}^s(\mathbb{R}^+) = \text{Ker } P$. $\square$

References

1. R.A. Adams, J.J. Fournier, J. F. John, *Sobolev Spaces* (Elsevier/Academic, Amsterdam, 2003)
2. S. Agmon, Spectral properties of Schrödinger operators and scattering theory. Ann. Scuola Norm. Sup. di Pisa **4**(2), 151–218 (1975)
3. S. Agmon, A. Duglis, L. Nirenberg, Estimates near the boundary for solutions of elliptic partial differential equations satisfying general boundary conditions. Commun. Pure Appl. Math. **12**(II), 623–727 (1959); **17** (1964) 35–92
4. V.M. Babich, M.A. Lyalinov, V.E. Grikurov, *The Sommerfeld-Malyuzhinets Technique in Diffraction Theory* (Alpha Science International, Oxford, 2007)
5. B. Baker, E. Copson, *The Mathematical Theory of Huygens Principle* (Clarendon Press, Oxford, 1939)
6. J.-M.L. Bernard, Diffraction by a metallic wedge covered with a dielectric material. Wave Motion **9**, 543–561 (1987)
7. J.-M.L. Bernard, A spectral approach for scattering by impedance polygons. Q. J. Mech. Appl. Math. **59**(4), 517–550 (2006)
8. L. Bers, F. John, M. Schechter, Partial differential equations, in *Lectures in Applied Mathematics*. Proceedings of the Summer Seminar, Boulder, CO, 1957 (Interscience Publishers/Wiley, New York, 1964)
9. A.-S. Bonnet-Bendhia, P. Joly, Mathematical analysis of guided water waves. SIAM J. Appl. Math. **53**(6), 1507–1550 (1993)
10. A.-S. Bonnet-Bendhia, M. Dauge, K. Ramdani, Spectral analysis and singularities of a non-coercive transmission problem. C. R. Acad. Sci., Paris, Sér. I, Math. **328**(8), 717–720 (1999)
11. M. Born, E. Wolf, *Principles of Optics* (Cambridge University Press, Cambridge, 1966)
12. V.A. Borovikov, *Diffraction by Polygons and Polyhedrons* (Nauka, Moscow, 1966, in Russian)
13. V.A. Borovikov, Diffraction by impedance wedge with curved faces. J. Commun. Technol. Electron. **43**(12), 25–27 (1998)
14. L. Boutet de Monvel, Comportement d'un operateur pseudo-differentiel sur une variete a bord. I: La propriete de transmission, II: Pseudo-noyaux de Poisson. J. Anal. Math. **17**, 241–253, 255–304 (1966)
15. F. Cajori, *A History of Physics* (Dover, New York, 1962)
16. H.S. Carslaw, Diffraction of waves by a wedge of any angle. Proc. Lond. Math. Soc. (Ser. 2) **18**(1), 291–306 (1920)
17. L.P. Castro, D. Kapanadze, Dirichlet indexDirichlet-Neumann impedance boundary-value problems arising in rectangular wedge diffraction problems. Proc. Am. Math. Soc. **136**, 2113–2123 (2008)

© Springer Nature Switzerland AG 2019

A. Komech, A. Merzon, *Stationary Diffraction by Wedges*, Lecture Notes
in Mathematics 2249, https://doi.org/10.1007/978-3-030-26699-8

18. L.P. Castro, D. Kapanadze, Wave diffraction by a 45 degree wedge sector with Dirichlet and Neumann boundary conditions. Math. Comput. Model. **48**(1/2), 114–121 (2008)
19. L.P. Castro, D. Kapanadze, Exterior wedge diffraction problems with Dirichlet, Neumann and impedance boundary conditions. Acta Appl. Math. **110**(1), 289–311 (2010)
20. L.P. Castro, D. Kapanadze, Wave diffraction by a 270 degrees wedge sector with Dirichlet, Neumann and impedance boundary conditions. Proc. A. Razmadze Math. Inst. **155**, 96–99 (2011)
21. L.P. Castro, D. Kapanadze, Wave diffraction by wedges having arbitrary aperture angle. Math. Methods Appl. Sci. **421**(2), 1295–1314 (2015)
22. L.P. Castro, F.-O. Speck, F.S. Teixeira, On a class of wedge diffraction problems posted by Erhard Meister. Oper. Theory Adv. Appl. **147**, 213–240 (2004)
23. L.P. Castro, F.-O. Speck, F.S. Teixeira, Mixed boundary value problems for the Helmholtz equation in a quadrant. Integr. Equ. Oper. Theory **56**, 1–44 (2006)
24. A. Choque, Yu. Karlovich, A. Merzon, P. Zhevandrov, On the convergence of the amplitude of the diffracted nonstationary wave in scattering by wedges. Russ. J. Math. Phys. **19**(3), 373–384 (2012)
25. P.C. Clemmow, *The Plane Wave Spectrum Representation of Electromagnetic Fields* (Pergamon Press, Oxford, 1966)
26. E.T. Copson, *Asymptotic Expansions* (Cambridge University Press, Cambridge, 1965)
27. M. Costabel, E. Stephan, Boundary integral equations for mixed boundary value problems in polygonal domains and Galerkin approximation. Banach Center Publ. **15**, 175–251 (1985)
28. R. Courant, D. Hilbert, *Methods of Mathematical Physics*, vol. I (Interscience Publishers, New York, 1953); vol. II, (Interscience Publishers, New York, 1962)
29. H. Crew (ed.), *The Wave Theory of Light*. Memoirs by Huygens, Young and Fresnel (American Book Company, Woodstock, 1900). https://archive.org/details/wavetheoryofligh00crewrich
30. M. Dauge, Elliptic boundary value problems on corner domains, in *Smoothness and Asymptotics of Solutions*. Lecture Notes in Mathematics, vol. 1341 (Springer, Berlin, 1988)
31. J.E. De la Paz Mendez, A.E. Merzon, Scattering of a plane wave by hard-soft wedges, in *Operator Theory: Advances and Applications*, vol. 220 (Birkhäuser/Springer Basel AG, Basel, 2012), pp. 207–227
32. P. Drude, *The Theory of Optics* (Dover, New York, 1959)
33. T. Ehrhardt, A.P. Nonalsco, F.-O. Speck, Boundary integral methods for wedge diffraction problems: the angle 2π. Dirichlet and Neumann conditions. Oper. Matrices **5**, 1–40 (2011)
34. T. Ehrhardt, A.P. Nonalsco, F.-O. Speck, A Riemann surface approach for diffraction from rational wedges. Oper. Matrices **8**, 301–355 (2014)
35. D.M. Eidus, The principle of limit amplitude. Russ. Math. Surv. **24**(3), 24–97 (1969)
36. G. Eskin, The wave equation in a wedge with general boundary conditions. Commun. Partial Differ. Equ. **17**, 99–160 (1992)
37. A. Esquivel, A.E. Merzon, Nonstationary scattering DN-problem of a plane wave by a wedge, in *Days on Diffraction. Proceedings of the International Conference*, St. Petersburg, 2006, pp. 187–196
38. A. Esquivel, A. Merzon, An explicit formula for the nonstationary diffracted wave scattered on a NN-wedge. Acta Apl. Math. **136**(1), 119–145 (2015)
39. D.V. Evans, N. Kuznetsov, Trapped modes, in *Gravity Waves in Water of Finite Depth*. Advances in Fluid Mecanics, vol. 10 (J.N.Hunt WIT Press, Southampton, 1997), pp. 127–168
40. A.F. Filippov, Study of the solution of the nonstationary problem of plane wave diffraction at an impedance wedge. USSR Comput. Math. Math. Phys. **7**(4), 143–156 (1967)
41. A.J. Fresnel, Note sur la Théorie de Diffraction, 1818, in *Oeuvres complètes de Augustin Fresnel*, ed. by H. de Senaramont, É. Verdet, L. Fresnel, vol. I (Imprimerie Impèrial, Paris, 1866, in French) (English translation: pp. 79–144 in Ref. [29])
42. P. Gérard, G. Lebeau, Diffusion d'une onde par un coin. J. Am. Math. Soc. **6**(2), 341–424 (1993)

43. F. M. Grimaldi, *Physico Mathesis de Lumine, Coloribus, et Iride, Aliisque Annexis Libri Duo* (Bologna ("Bonomia"), Italy/Vittorio Bonati, 1665), pp. 1–11. [Latin] https://books.google. at/books?id=FzYVAAAAQAAJ&pg=PA1&redir$_$esc=y#v=onepage&q&f=false

44. P. Grisvard, *Elliptic Problems in Nonsmooth Domains* (Pitman, Boston, 1985)

45. D.P. Hewett, J.R. Ockendon, D.J. Allwright, Switching on a two-dimensional time-harmonic scalar wave in the presence of a diffracting edge. Wave Motion **48**(3), 197–213 (2011)

46. L. Hörmander, *The Analysis of Linear Partial Differential Operators I: Distribution Theory and Fourier Analysis* (Springer, Berlin, 2003)

47. H.C. Huygens, *Traité de la lumiere, Pierre vander Aa, marchand libraire, Leide* (1690, in French). https://archive.org/details/bub_gb_kVxsaYdZaaoC (English translation: pp. 1–41 in Ref. [29])

48. W. Ignatowsky, Reflexion elektromagnetischer Wellen an einem Drahte. Ann. der Physik **18**(13), 495–522 (1905)

49. A. Jensen, T. Kato, Spectral properties of Schrödinger operators and time-decay of the wave functions. Duke Math. J. **46**, 583–611 (1979)

50. I. Kay, The diffraction of an arbitrary pulse by a wedge. Commun. Pure Appl. Math. **6**, 521–546 (1953)

51. J. Keller, A. Blank, Diffraction and reflection of pulses by wedges and corners. Commun. Pure Appl. Math. **4**(1), 75–95 (1951)

52. A.I. Komech, Elliptic boundary value problems on manifolds with piecewise smooth boundary. Math. USSR Sbornik **21**(1), 91–135 (1973)

53. A.I. Komech, Elliptic differential equations with constant coefficients in a cone. Moscow Univ. Math. Bull. Math. Mech. Ser. **29**(2), 140–145 (1974)

54. A.I. Komech, Linear partial differential equations with constant coefficients, in *Partial Differential Equations II (Encyclopaedia of Mathematical Sciences)*, vol. 31 (Springer, Berlin, 1999), pp.127–260

55. A.I. Komech, *Quantum Mechanics. Genesis and Achievements* (Springer, Dordrecht, 2013)

56. A.I. Komech, On dynamical justification of quantum scattering cross section. J. Math. Anal. Appl. **432**(1), 583–602 (2015). ArXiv:1206.3677

57. A.I. Komech, E.A. Kopylova, *Dispersion Decay and Scattering Theory* (Wiley, Hoboken, 2012)

58. A.I. Komech, A.E. Merzon, General boundary value problems in region with corners, in *Operator Theory. Advances and Applications*, vol. 57 (Birkhauser, Basel, 1992), pp. 171–183

59. A.I. Komech, A.E. Merzon, Sommerfeld representation in stationary diffraction problems in angles, in *Spectral and Evolutional Problems*, vol. 4 (Simferopol State University, Simferopol, 1995), pp. 210–219

60. A.I. Komech, A.E. Merzon, Limiting amplitude principle in the scattering by wedges. Math. Methods Appl. Sci. **29**(10), 1147–1185 (2006)

61. A.I. Komech, A.E. Merzon, Relation between Cauchy data for the scattering by a wedge. Russ. J. Math. Phys. **14**(3), 279–303 (2007)

62. A.I. Komech, A.E. Merzon, On uniqueness and stability of Sobolev's solution in scattering by wedges. J. Appl. Math. Phys. (ZAMP) **66**(5), 2485–2498 (2015)

63. A.I. Komech, A.E. Merzon, P.N. Zhevandrov, On completeness of Ursell's trapping modes. Russ. J. Math. Phys. **4**(4), 55–85 (1996)

64. A. Komech, A. Merzon, P. Zhevandrov, A method of complex characteristics for elliptic problems in angles, and its applications, in *Partial Differential Equations*. American Mathematical Society Translations, vol. 206(2) (American Mathematical Society, Providence, 2002), pp. 125–159

65. A.I. Komech, N.J. Mauser, A.E. Merzon, On Sommerfeld representation and uniqueness in scattering by wedges. Math. Methods Appl. Sci. **28**(2), 147–183 (2005)

66. A.I. Komech, A.E. Merzon, J.E. De La Paz Méndez, Time-dependent scattering of generalized plane waves by wedges. Math. Methods Appl. Sci. **38**(18), 4774–4785 (2015)

67. A.I. Komech, A.E. Merzon, A. Esquivel Navarrete, J.E. De La Paz Méndez, T.J. Villalba Vega, Sommerfeld's solution as the limiting amplitude and asymptotics for narrow wedges. Math. Methods Appl. Sci., 1–14 (2018). https://doi.org/10.1002/mma.5075

68. V.A. Kondratiev, Boundary value problems for elliptic equations in domains with conical or angular points. Trans. Moscow Math. Soc. **16**, 227–313 (1967)

69. V.A. Kondratiev, The smoothness of a solution of Dirichlet's problem for 2nd order elliptic equations in a region with a piecewise smooth boundary. Diff. Equ. **6**(10), 1392–1401 (1970)

70. N.G. Kuznetssov, V.G. Maz'ya, B.R.Vainberg, *Lectures on Linear Time-Harmonic Water Waves* (Department of Mathematics, Technical Report Series, UNC Charlotte, 2000), pp. 28223

71. O. Lafitte, *The Wave Diffracted by a Wedge with Mixed Boundary Conditions* (Société Mathématique de France, Paris, 2002)

72. M.A. Leontovich, On the approximate boundary conditions for the electromagnetic field on the surface of well conducting bodies, in *Investigations of Radio Waves Propagation*, ed. by B.A. Vvedensky (Academy of Sciences of USSR, Moscow, 1948, in Russian), pp. 5–12

73. R.S. Lewy, H. Lewy, Uniqueness of water waves on a sloping beach. Commun. Pure Appl. Math. **14**, 521–546 (1961)

74. G. Libri, *Histoire des Sciences Mathématiques en Italie*, vol. III (Renouard, Paris, 1840)

75. H.S. Lipson, *The Great Experiments in Physics* (Oliver & Boyd, Edinburg, 1968)

76. G.D. Malujinetz, Excitation, reflection and emission of surface waves from a wedge with given face impedances. Sov. Phys. Dokl. **3**, 752–755 (1959)

77. V.A. Malyshev, *Random Walks, Wiener-Hopf Equations in the Quadrant of Plane, Galois Automorphisms* (Moscow University, Moscow, 1970, in Russian)

78. G.D. Malyuzhinets, Inversion formula for the Sommerfeld integral. Sov. Phys. Dokl. **3**, 52–56 (1958)

79. J.C. Maxwell, *A Treatise on Electricity and Magnetism* (Clarendon Press, Oxford, 1873)

80. V.G. Maz'ya, B.A. Plamenevskii, Problems with oblique derivatives in regions with piecewise smooth boundaries. Funct. Anal. Appl. **5**, 256–258 (1971)

81. V.G. Maz'ya, B.A. Plamenevskii, On boundary value problems for a second-order elliptic equations in a domain with wedges. Vestn. Leningr. Univ. Math. **1**, 102–108 (1975)

82. E. Meister, Some solved and unsolved canonical problems of diffraction theory, in *Lecture Notes in Mathematical*, vol. 1285 (Springer, Berlin, 1987), pp. 320–336

83. E. Meister, F.O. Speck, F.S. Teixeira, Wiener-Hopf-Hankel operators for some wedge diffraction problems with mixed boundary conditions. J. Integr. Equ. Appl. **4**(2), 229–255 (1992)

84. E. Meister, F. Penzel, F.O. Speck, F.S. Teixeira, Some interior and exterior boundary-value problems for the Helmholtz equations in a quadrant. Proc. R. Soc. Edinburgh Sect. A **123**(2), 275–294 (1993)

85. E. Meister, F. Penzel, F.-O. Speck, F.S. Teixeira, Two canonical wedge problems for the Helmholtz equation. Math. Methods Appl. Sci. **17**, 877–899 (1994)

86. E. Meister, A. Passow, K. Rottbrand, New results on wave diffraction by canonical obstacles. Oper. Theory Adv. Appl. **110**, 235–256 (1999)

87. A.E. Merzon, On the solvability of differential equations with constant coefficients in a cone. Sov. Math. Dokl. **14**(4), 1012–1015 (1973)

88. A.E. Merzon, General boundary value problems for the Helmholtz equations in a plane angle. Uspekhi Math. Nauk. **32**(2), 219–220 (1977, in Russian)

89. A.E. Merzon, On Ursell's problem, in *Proceedings of the Third International Conference on Mathematical and Numerical Aspects of Wave Propagation*, ed., by G. Cohen. SIAM-INRIA (1995), pp. 613–623

90. A.E. Merzon, Sommerfeld solution as limiting amplitude diffraction by wedge, in *Abstracts Sommerfeld '96—Workshop, Modern Mathematical Methods in Diffraction Theory and Its Applications in Engineering*, ed. by E. Meister. Preprint Nr.1868 (Technische Hochschule Darmstadt, Darmstadt, 1996), p. 29

91. A.E. Merzon, Well-posedness of the problem of nonstationary diffraction of Sommerfeld, in *Proceedings of the International Conference Day on Diffraction 2003* (University of St. Petersburg, St. Petersburg, 2003), pp. 151–162

92. A.E. Merzon, J.E. De La Paz Méndez, DN-Scattering of a plane wave by wedges. Math. Methods Appl. Sci. **34**(15), 1843–1872 (2011)

93. A.E. Merzon, P. Zhevandrov, Asymptotics of edge waves on a beach of nonconstant slope. Uspekhi Math. Nauk. **51**(5), 220 (1996, in Russian)

94. A.E. Merzon, P.N. Zhevandrov, High-frequency asymptotics of edge waves on a beach of nonconstant slope. SIAM J. Appl. Math. **59**(2), 529–546 (1998)

95. A.E. Merzon, P.N. Zhevandrov, On the Neumann problem for the Helmholtz equation in a plane angle. Math. Meth. Appl. Sci. **23**(16), 1401–1446 (2000)

96. A.E. Merzon, F.O. Speck, T.J. Villalba, On the weak solution of the Neumann problem for the 2D Helmholtz equation in a convex cone and H^s regularity. Math. Meth. Appl. Sci. **34**(1), 24–43 (2011)

97. A.E. Merzon, A.I. Komech, J.E. De la Paz Mendez, T.J. Villalba, On the Keller-Blank solution to the scattering problem of pulses by wedges. Math. Methods Appl. Sci. **38**(10), 2035–2040 (2015)

98. A.E. Merzon, P.N. Zhevandrov, J.E. De La Paz Mendez, On the behavior of the edge diffracted nonstationary wave in scattering by wedges near the front. Russ. J. Math. Phys. **22**(4), 491–503 (2015)

99. C. Morawetz, The limiting amplitude principle. Commun. Pure Appl. Math. **15**, 349–361 (1962)

100. R.J. Nagem, M. Zampolli, G. Sandri, *Arnold Sommerfeld. Mathematicai Theory of Diffraction, Progress in Mathematical Physics*, vol. 35 (Springer Science+Business Media, New York, 2004) Originally published by Birkhauser Boston in 2004

101. S.A. Nazarov, B.A. Plamenevskij, *Elliptic Problems in Domains with Piecewise Smooth Boundaries* (De Gruyter, Berlin, 1994)

102. I. Newton, *The Principia: The Authoritative Translation and Guide: Mathematical Principles of Natural Philosophy*. A new translation by I.B. Cohen, A. Whitman, Preceded by A Guide to Newton's Principia by I.B. Cohen (University of California Press, Berkeley, 1999)

103. A.P. Nolasco, F.-O. Speck, On some boundary value problems for Helmholtz equation in cone 240°, in *Operator Theory: Advances and Applications*, vol. 221 (Birkhauser, Basel, 2012), pp. 497–513

104. F. Oberhettinger, Diffraction of waves by a wedge. Commun. Pure Appl. Math. **7**, 551–563 (1954)

105. F. Oberhettinger, On asymptotic series for functions occurring in the theory of diffraction of waves by wedges. J. Math. Phys. **34**(4), 245–255 (1956)

106. F. Oberhettinger, On the diffraction and reflection of waves and pulses by wedges and corners. J. Res. Natl. Bur. Stand. **61**(2), 343–365 (1958)

107. A.V. Osipov, Diffraction by a wedge with higher-order boundary conditions. Radio Sci. **31**(6), 1705–1720 (1996)

108. V.M. Papadopoulos, Pulse diffraction by an imperfectly reflecting wedge. J. Austral. Math. Soc. **2**(1), 97–106 (1961)

109. W. Pauli, On asymptotic series for functions in the theory of diffraction of light. Phys. Rev. **54**, 924–931 (1938)

110. F. Penzel, F.S. Teixeira, The Helmholtz equation in a quadrant with Robin's conditions. Math. Methods Appl. Sci. **22**, 201–216 (1999)

111. A.S. Peters, Water waves over sloping beaches and the solution of a mixed boundary value problem for $\Delta\Phi - k^2\Phi = 0$ in a sector. Commun. Pure Appl. Math. **5**, 87–108 (1952)

112. G.I. Petrashen', V.G. Nikolaev, D.P. Kouzov, On the method of series in the theory of diffraction of waves by plane corner regions. Sci. Notes LGU **246**(5), 5–70 (1958, in Russian)

113. J.H. Poincaré, Sur la polarization par diffraction. Acta Math. **16**, 297–339 (1892)

114. J.H. Poincaré, Sur la polarization par diffraction. Acta Math. **20**, 313–355 (1897)

115. M. Reed, B. Simon, *Methods of Modern Mathematical Physics II: Fourier Analysis, Self-Adjointness* (Academic, New York, 1975)

116. M. Roseau, Short waves parallel to the shore over a sloping beach. Commun. Pure Appl. Math. **11**, 433–493 (1958)

117. K. Rottbrand, Time-dependent plane wave diffraction by a half-plane: explicit solution for Rawlins' mixed initial boundary value problem. Z. Angew. Math. Mech. **78**(5), 321–335 (1998)

118. K. Rottbrand, *Exact Solution for Time-Dependent Diffraction of Plane Waves by Semi-Infinite Soft/Hard Wedges and Half-planes.* Preprint Technical University Darmstadt, no. 1984 (1998)

119. M.P. Sakharova, A.F. Filippov, The solution of non-stationary problem of the diffraction at an impedance wedge using tabulated functions. U.S.S.R. Comput. Math. Math. Phys. **7**(3), 128–144 (1967)

120. I.V. Savelyev, Physics, A general course, in *Electricity and Magnetism, Waves and Optics* vol. 2 (Mir, Moscow, 1989)

121. S.H. Schot, Eighty years of Sommerfeld's radiation condition. Hist. Math. **19**(4), 385–401 (1992)

122. L. Schwartz, *Théorie Des Distributions* (Hermann, Paris, 1966, in French)

123. G.E. Shilov, On boundary value problems in a quarter plane for partial differential equations with constant coefficients. Sib. Math. J. **11**(1), 144–160 (1961, in Russian)

124. V.I. Smirnov, S.L. Sobolev, Sur une méthode nouvelle dans le probléme plan des vibrations élastiques. Trudy Seismol. Inst. Acad. Nauk SSSR **20**, 1–37 (1932)

125. S. L. Sobolev, Theory of diffraction of plane waves, in *Proceedings of Seismological Institute*, vol. 41 (Russian Academy of Science, Leningrad, 1934)

126. S.L. Sobolev, General theory of diffraction of waves on Riemann surfaces, Tr. Fiz.-Mat. Inst. Steklova **9**, 39–105 (1935, in Russian) (English translation: S.L. Sobolev, General theory of diffraction of waves on Riemann surfaces, pp. 201–262 in: Selected Works of S.L. Sobolev, vol. I, Springer, New York, 2006)

127. S.L. Sobolev, Some questions in the theory of propagations of oscillations, in *Differential and Integral Equations of Mathematical Physics, Leningrad-Moscow*, ed. by F. Frank, P. Mizes (1937, in Russian), pp. 468–617

128. S.L. Sobolev, On mixed problem for partial differential equations with two independent variables. Dokl. Ac. Sci. USSR **122**(4), 555–558 (1958, in Russian)

129. A. Sommerfeld, Mathematische theorie der diffraction. Math. Ann. **47**, 317–341 (1896)

130. A. Sommerfeld, Theoretisches über die Beugung der Röntgenstrahlen (German). Z. Math. Phys. **46**, 11–97 (1901)

131. A. Sommerfeld, Die Greensche Funktion der Schwingungsgleichung. Thematiker-Vereinigung **21**, 309–353 (1912). Reprinted in Gesammelte Schriften, vol. 1, pp. 272–316

132. A. Sommerfeld, Autobiographische Skizze, in *Gesammelte Schriften*, vol. 4 (1951), pp. 673–682

133. A. Sommerfeld, *Lectures on Theoretical Physics*, vol. 4 (Optics, New York, 1954)

134. E.M. Stein, Note on singular integrals. Proc. Am. Math. Soc. **8**, 250–254 (1957)

135. G.G. Stokes, Report on recent researches in hydrodynamics. Rep. Br. Assoc. **1**(1), 157–187 (1846), in *Mathematical and Physical Papers*, ed. by G.G. Stokes, vol. I (Cambridge University Press, Cambridge, 1880)

136. A.N. Tikhonov, A.A. Samarskii, On principle of radiation. JETP **18**(2), 243–248 (1948)

137. A.A.Tuzilin, Diffraction of a plane sonic wave in an angular domain whose boundaries are absolutely rigid and slippery and coated thin elastic plates. (Russian) Differ. Uravn. **9**, 1875–1888 (1973)

138. A.A.Tuzilin, On the theory of Maljuinec's inhomogeneous functional equations. (Russian) Differ. Uravn. **9**, 2058–2064 (1973)

139. F. Ursell, Edge waves on a sloping beach. Proc. R. Soc. Lond. **A214**, 79–97 (1952)

140. B.R. Vainberg, *Asymptotic Methods in Equations of Mathematical Physics* (Gordon and Breach, New York, 1989)

141. M.I. Vishik, G.I. Eskin, Equations in convolutions in a bounded region. Russ. Math. Surv. **20**(3), 85–151 (1965)

142. G.N. Watson, *A Treatise on the Theory of Bessel Functions* (Cambridge University Press, Cambridge, 1966)

143. W.L. Wendland, E. Stephan, G.C. Hsiao, On the integral equation method for the plane mixed boundary value problem of the Laplacian. Math. Methods Appl. Sci. **1**(3), 265–321 (1979)
144. E.L. Whittaker, *A History of the Theories of Aether and Electricity*. The Classical Theories, vol. 1 (Longmans, London, 1910) (reprinted by Oxford City Press, 2012)
145. E.L. Whittaker, G.N. Watson, *A Course of Modern Analysis* (Macmillan, New York, 1948)
146. T. Young, Experiments and calculations relative to the physical optics. A Bakerian lecture. Read November 24, 1803. Philos. Trans. R. Soc. Lond. Part London (1804) (68–76 in Ref. [29])
147. T. Young, in *Miscellaneous Works*, ed. by G. Peacock, J. Leitch, vol. I (John Murray, London, 1855). https://archive.org/stream/miscellaneouswo01youngoog#page/n11/mode/2up/
148. P.N. Zhevandrov, A.E. Merzon, Stability of entrained surface waves under small perturbations of the density of the upper layer. Math. Notes **64**, 814–817 (1998)
149. P. Zhevandrov, A. Merzon, On the Neumann Problem for the Helmholtz equation in a plane angle. Math. Methods Appl. Sci. **23**, 1401–1446 (2000)
150. E.I. Zverovich, G.S. Litvinchuk, Boundary-value problems with a shift for analytic functions and singular functional equations. Russ. Math. Surv. **23**(3), 67–124 (1968)

Index

© Springer Nature Switzerland AG 2019

A. Komech, A. Merzon, *Stationary Diffraction by Wedges*, Lecture Notes in Mathematics 2249, https://doi.org/10.1007/978-3-030-26699-8

LECTURE NOTES IN MATHEMATICS

Editors in Chief: J.-M. Morel, B. Teissier;

Editorial Policy

1. Lecture Notes aim to report new developments in all areas of mathematics and their applications – quickly, informally and at a high level. Mathematical texts analysing new developments in modelling and numerical simulation are welcome.

   Manuscripts should be reasonably self-contained and rounded off. Thus they may, and often will, present not only results of the author but also related work by other people. They may be based on specialised lecture courses. Furthermore, the manuscripts should provide sufficient motivation, examples and applications. This clearly distinguishes Lecture Notes from journal articles or technical reports which normally are very concise. Articles intended for a journal but too long to be accepted by most journals, usually do not have this "lecture notes" character. For similar reasons it is unusual for doctoral theses to be accepted for the Lecture Notes series, though habilitation theses may be appropriate.

2. Besides monographs, multi-author manuscripts resulting from SUMMER SCHOOLS or similar INTENSIVE COURSES are welcome, provided their objective was held to present an active mathematical topic to an audience at the beginning or intermediate graduate level (a list of participants should be provided).

   The resulting manuscript should not be just a collection of course notes, but should require advance planning and coordination among the main lecturers. The subject matter should dictate the structure of the book. This structure should be motivated and explained in a scientific introduction, and the notation, references, index and formulation of results should be, if possible, unified by the editors. Each contribution should have an abstract and an introduction referring to the other contributions. In other words, more preparatory work must go into a multi-authored volume than simply assembling a disparate collection of papers, communicated at the event.

3. Manuscripts should be submitted either online at www.editorialmanager.com/lnm to Springer's mathematics editorial in Heidelberg, or electronically to one of the series editors. Authors should be aware that incomplete or insufficiently close-to-final manuscripts almost always result in longer refereeing times and nevertheless unclear referees' recommendations, making further refereeing of a final draft necessary. The strict minimum amount of material that will be considered should include a detailed outline describing the planned contents of each chapter, a bibliography and several sample chapters. Parallel submission of a manuscript to another publisher while under consideration for LNM is not acceptable and can lead to rejection.

4. In general, **monographs** will be sent out to at least 2 external referees for evaluation.

   A final decision to publish can be made only on the basis of the complete manuscript, however a refereeing process leading to a preliminary decision can be based on a pre-final or incomplete manuscript.

   Volume Editors of **multi-author works** are expected to arrange for the refereeing, to the usual scientific standards, of the individual contributions. If the resulting reports can be forwarded to the LNM Editorial Board, this is very helpful. If no reports are forwarded or if other questions remain unclear in respect of homogeneity etc, the series editors may wish to consult external referees for an overall evaluation of the volume.

5. Manuscripts should in general be submitted in English. Final manuscripts should contain at least 100 pages of mathematical text and should always include

   - a table of contents;
   - an informative introduction, with adequate motivation and perhaps some historical remarks: it should be accessible to a reader not intimately familiar with the topic treated;
   - a subject index: as a rule this is genuinely helpful for the reader.
   - For evaluation purposes, manuscripts should be submitted as pdf files.

6. Careful preparation of the manuscripts will help keep production time short besides ensuring satisfactory appearance of the finished book in print and online. After acceptance of the manuscript authors will be asked to prepare the final LaTeX source files (see LaTeX templates online: https://www.springer.com/gb/authors-editors/book-authors-editors/manuscriptpreparation/5636) plus the corresponding pdf- or zipped ps-file. The LaTeX source files are essential for producing the full-text online version of the book, see http://link.springer.com/bookseries/304 for the existing online volumes of LNM). The technical production of a Lecture Notes volume takes approximately 12 weeks. Additional instructions, if necessary, are available on request from lnm@springer.com.

7. Authors receive a total of 30 free copies of their volume and free access to their book on SpringerLink, but no royalties. They are entitled to a discount of 33.3 % on the price of Springer books purchased for their personal use, if ordering directly from Springer.

8. Commitment to publish is made by a *Publishing Agreement*; contributing authors of multiauthor books are requested to sign a *Consent to Publish form*. Springer-Verlag registers the copyright for each volume. Authors are free to reuse material contained in their LNM volumes in later publications: a brief written (or e-mail) request for formal permission is sufficient.

Addresses:
Professor Jean-Michel Morel, CMLA, École Normale Supérieure de Cachan, France
E-mail: moreljeanmichel@gmail.com

Professor Bernard Teissier, Equipe Géométrie et Dynamique,
Institut de Mathématiques de Jussieu – Paris Rive Gauche, Paris, France
E-mail: bernard.teissier@imj-prg.fr

Springer: Ute McCrory, Mathematics, Heidelberg, Germany,
E-mail: lnm@springer.com

Printed in the United States
By Bookmasters